Ivy Becker

Structural and diagenetic controls on reservoir quality in tight siliciclastic and carbonate rocks

Structural and diagenetic controls on reservoir quality in tight siliciclastic and carbonate rocks

by
Ivy Becker

KIT Scientific Publishing

Karlsruher Institut für Technologie
Institut für Angewandte Geowissenschaften

Structural and diagenetic controls on reservoir quality
in tight siliciclastic and carbonate rocks

Zur Erlangung des akademischen Grades eines Doktors der Naturwissenschaften von der KIT-Fakultät für Bauingenieur-, Geo- und Umweltwissenschaften des Karlsruher Instituts für Technologie (KIT) genehmigte Dissertation

von Ivy Becker aus Hohenlimburg, Stadt Hagen

Tag der mündlichen Prüfung: 01. August 2018
Erster Gutachter: Prof. Dr. Christoph Hilgers
Zweiter Gutachter: Prof. Dr. Reinhard Gaupp
Dritter Gutachter: Prof. Stephen E. Laubach, PhD

Impressum

KIT Scientific Publishing

Karlsruher Institut für Technologie (KIT)
KIT Scientific Publishing
Straße am Forum 2
D-76131 Karlsruhe

KIT Scientific Publishing is a registered trademark
of Karlsruhe Institute of Technology.
Reprint using the book cover is not allowed.

www.ksp.kit.edu

Print on Demand 2018 – Gedruckt auf FSC-zertifiziertem Papier

ISBN 978-3-7315-0836-6
DOI 10.5445/KSP/1000085503

Structural and diagenetic controls on reservoir quality in tight siliciclastic and carbonate rocks

zur Erlangung des akademischen Grades eines

Doktors der Naturwissenschaften

von der Fakultät für Bauingenieur-, Geo- und Umweltwissenschaften
des Karlsruher Instituts für Technologie (KIT)

genehmigte

Dissertation

von

Ivy Becker

aus Hohenlimburg, Stadt Hagen

Tag der mündlichen Prüfung: 01. August 2018
Erster Gutachter: Prof. Dr. Christoph Hilgers
Zweiter Gutachter: Prof. Dr. Reinhard Gaupp
Dritter Gutachter: Prof. Stephen E. Laubach, PhD

Acknowledgements

My sincerest thanks to Christoph Hilgers who gave me the chance to new extensive experiences by moving to Karlsruhe and entrusting me with the management of the projects that are the basis of this work. His support and guidance, and most importantly his enthusiasm for new concepts and methods greatly improved the results and constantly lead to new, motivating ideas for future tasks. Thank you for countless long, and often late-time discussions that were an essential part of this work.

Reinhard Gaupp and Stephen Ernest Laubach are thankfully acknowledged for co-refereeing my work. Many thanks for helpful discussions and comments that greatly improved this thesis.

Bastian Koehrer is gratefully acknowledged for his time reviewing manuscripts or conference presentations and -abstracts, and countless helpful ideas and comments which greatly improved the applied part of this work. His effort in spreading the results by setting up internal Wintershall meetings and introducing me to staff played a crucial role in advancing the projects.

Additional support by Dirk Adelmann, his time on reviewing manuscripts, introducing me to Wintershall staff, and fruitful status or conference meetings are thankfully acknowledged.

Further discussions with or reviews by Frauke Schulze, Gregor Hollmann, Wolfgang Jelinek, Claas Mennerich, Jan Himmerkus, Frauke Schäfer, Dominik Weiel, and Michael Peter Süss concerning the provision of data, workflows, or the general progress of the project are greatly acknowledged. Wintershall Holding GmbH is acknowledged for funding this project and the provision of data. Together with their license partners ExxonMobil Production Deutschland GmbH and von Rautenkranz Exploration und Produktion GmbH, they are kindly thanked for the permission to publish the data.

Norbert Schleifer, Philipp Aurin (both Wintershall Holding GmbH), Bernhard Krooss, and Alexandra Amann-Hildenbrand (both RWTH Aachen University) are sincerely thanked for their support and time for countless discussions on the set-up of the petrophysical measurements, which greatly improved the results of this work.

Additional helpful discussions, ideas, and reviews by Wolf-Dieter Karnin on various topics concerning the German sedimentary hydrocarbon reservoirs and his willingness to share his wide-spread experiences, both in the industry and academia, are gratefully acknowledged.

Marita Felder and Nicolaas Molenaar are thankfully acknowledged for their time discussing sample preparation and diagenesis and sharing an interesting field trip to the Upper Carboniferous outcrops of this study. Thorough manuscript reviews and helpful comments by Marita Felder greatly improved the results and are gratefully acknowledged.

The Natursteinbetrieb Woitzel GmbH, CEMEX Kies Split GmbH, Rump und Salzmann GmbH, Lhoist Germany – Rheinkalk GmbH, and Fels Werke GmbH are kindly thanked for granting access to their quarries, pleasant and safe working conditions, and the permission to publish the data.

Birgit Müller, Jörg Meixner, and Patrick Wüstefeld are gratefully thanked for helpful discussions and support with modeling issues. The provision of academic licenses to their software by BeicipFranlab, Schlumberger, and 3DStress by Alan Morris is thankfully acknowledged.

Sample preparation by Thomas Derichs (RWTH Aachen University) and Stephan Unrein (KIT Karlsruhe Institute of Technology) and their continuous motivation to improve the workflow is highly appreciated.

Many thanks to my colleague Tina Schmidt for taking the risk of a new start in Karlsruhe together with me and for many challenging but pleasant and successful field trips to collect the data that enabled this work.

Benjamin Busch is thankfully acknowledged for his countless hours discussing and reviewing new ideas and models and for his input that generally improved the results of this study. Many thanks for making the field trips with our students to England and France as pleasant as they were.

My thanks to all students who wrote their final theses supervised by me and to all student researchers I was working with. Marius Waldvogel and Simon Freitag are thanked for numerous fracture measurements and sample collection in the Zechstein outcrops that helped me generate a large data set. Sophie Behrendsen supported the results of this study by providing additional gamma ray measurements of the Zechstein carbonates. Data and sample collection during joint field trips to the Upper Carboniferous sandstones by Daniel Bücken are thankfully acknowledged.

Finally, my family and friends are sincerely thanked for their support and understanding which granted me the time and space to accomplish this thesis.

Abstract

Upper Carboniferous (Westphalian C/D) fluvial sandstones and Zechstein Ca2 (Stassfurt, second cycle) carbonates represent two important hydrocarbon reservoir units in NW Europe. Understanding their spatial variability of reservoir properties is hence key to reduce uncertainties in reservoir characterization of those tight, gas-producing formations. This study focuses on the comparison of Upper Carboniferous and Zechstein Ca2 outcrop analogs and reservoirs in NW Germany as outcrops provide supplementary information about the lateral variability of reservoir characteristics.

In the tight Upper Carboniferous fluvial sandstones, a better understanding of reservoir quality variations is crucial to develop successful exploration strategies.

Two outcrops of Upper Carboniferous (Westphalian D) sandstones are studied in the southern part of the Lower Saxony Basin, NW Germany, which can be regarded as analogs for gas-producing reservoirs in the north of the basin. The lower Westphalian D Woitzel quarry is separated by a lateral distance of approximately 15 km and a vertically stratigraphic distance of approximately 600 m to the upper Westphalian D Hüggel outcrop.

Better reservoir qualities of the upper Westphalian D Hüggel sandstones compared to the lower Westphalian D Woitzel sandstones are associated with the presence of authigenic kaolinite in the Hüggel samples. In contrast, reservoir quality deterioration of the Woitzel samples is due to intense illitization processes. For samples with permeabilities above 0.01 mD a positive correlation between permeability and average grain size is shown. Permeability variations are further related to sedimentary geometry of the sandstones as differences of two orders of magnitude from 0.001 to 0.1 mD occur within a single sandstone unit over a lateral distance of 150 m. These lateral variations in reservoir flow characteristics are influenced by repeated, centimeter-scaled fining-upwards grading and bed-internal erosion.

Results are integrated with data of a third outcrop (Piesberg) and two gas fields (drilled by Well A and Well B), all comprising of Westphalian C/D strata. Their petrophysical data sets and paragenetic sequences related to the burial- and structural history of the Lower Saxony Basin are used to derive controlling factors on reservoir quality on a basin-scale.

The Woitzel and Hüggel outcrop samples and the subsurface sandstones show similar matrix permeabilities but matrix porosities vary. In similitude, differences occur in their burial histories. Basin inversion resulted in stronger uplift of field B compared to field A reducing the formation potential of authigenic quartz and therefore results in higher matrix porosities in field B. With regard to the outcrop sandstones, further increased matrix porosities are related to carbonate cement dissolution in the late stages of uplift. In contrast, significantly reduced reservoir quality resulting from intense additional mineral cementation is present in the Piesberg sandstones, which is associated with temperature anomalies in the vicinity of faults. However, faulting and natural fracturing might also enhance the reservoir characteristics of field B. Such structures can act as potential migration pathways for leaching fluids or the production potential is increased by fracture permeability.

Integrated studies of the depositional setting, burial related diagenetic processes, and structural characteristics provide the foundation for understanding the heterogeneities in tight fluvial sandstones and thus, for future exploration and appraisal efforts in the Lower Saxony Basin and elsewhere.

In the Zechstein Ca2 carbonates, a better understanding of the geometries of natural fracture networks controlling the hydrocarbon flow paths in tight reservoirs is essential to improve reservoir modeling and well planning.

As variations in reservoir properties are related to different depositional environment types, understanding the lateral distribution of carbonate facies is key to reduce uncertainties in reservoir characterization. Three outcrops of Zechstein Ca2 carbonates along the SW margin of the Harz mountains are in the focus of this study. Their deposition took place at the southern margin of the Southern Permian Basin on the Eichsfeld-Altmark Swell, a NNE – SSW striking paleo-high.

Different gross depositional environments from slope to platform margin and inner platform deposition are interpreted for each outcrop. Based on that classification, the outcrops are correlated across the Eichsfeld-Altmark Swell over a lateral distance of 25 km. Reservoir properties are associated with the interpreted depositional environment type and compared to subsurface data of a gas-producing Ca2 field, which is located at a distance of 130 km to the NW. The provided porosity values of a carbonate slope reservoir are similar to the outcrop slope carbonates, but permeability values are increased compared to the outcrop values. Best reservoir qualities are related to carbonates from the platform margin, which show highest proportions in the center of the studied paleo-high. Information about the depositional environment distribution in the three studied outcrops can be used to choose a suitable outcrop analog to compare the outcrop fracture characteristics to subsurface data.

The Ührde quarry is chosen as analog for a studied horizontal well, that penetrated a Ca2 carbonate reservoir in NW Germany. Both contain dolomite as main lithology, which was deposited on a similar carbonate platform slope environment and overprinted by similar diagenetic and tectonic events.

Fracture characteristics of the outcrop and reservoir are compared, both indicating W – E orientations for well path planning due to dominantly northerly open fracture strike orientations. Outcrop fracture data are measured with a terrestrial laser scanner (t-LiDAR) in combination with a novel workflow for fracture surface detection firstly applied to carbonate rocks. Results are successfully validated using conventional compass measurements. Fracture intensities (P10 values describing the amount of fractures intersecting the scanline or well) are shown to be similar for the digital outcrop- and subsurface fractures, but both appear to be lower than manually measured values. This indicates a bias due to under-detected fractures oriented perpendicular to the wall in the t-LiDAR data set and a similar bias due to underrepresented fracture paralleling the well in the subsurface data set.

In addition to understanding the orientation, spacing, and connectivity of the fracture network in the Ca2 carbonate reservoirs, incorporation of the present-day stress field controlling the actual opening potential of the fractures at depth is crucial. Dilation- and slip tendency analyses provide a means to improve the fracture permeability estimations.

Fracture data are obtained from resistivity image log data of the studied horizontal well drilling a Ca2 carbonate reservoir in the Lower Saxony Basin. The dilation- and slip tendencies of those fractures are calculated to quantify their potential to contribute to the flow within the reservoir. Calculations are based on two different stress scenarios that are defined in accordance to stress data of a nearby field, where the orientation of the principal horizontal stress varies above and below the Zechstein salt, acting as a detachment.

Sub-vertical fractures show the highest potential to dilate and fractures with dip angles of 60° are favorably oriented for slip. Highest dilation- and slip tendencies are calculated for fractures paralleling the orientation of the principal horizontal stress, and hence, results depend on the strike and dip angle of the fractures. The opening potential of a fracture controls the fracture permeability and thus, incorporating stress field data significantly improves the calculation of the actual flow potential of a reservoir.

Kurzfassung

Fluviatile Sandsteine des Oberkarbons (Westphal C/D) und Karbonate des Zechstein Ca2 (Stassfurt, zweiter Zyklus) repräsentieren wichtige Kohlenwasserstoffreservoireinheiten in NW Europa. Deshalb ist die Kenntnis möglicher räumlicher Variationen ihrer Reservoireigenschaften ein wichtiger Schlüssel zur Eingrenzung von Unsicherheiten in der Charakterisierung dieser Lagerstätten. Diese Arbeit behandelt den Vergleich von Aufschlussanalog- und Reservoirgesteinen des Oberkarbons und Zechstein Ca2 in NW Deutschland und stellt dabei den zusätzlichen Wert lateraler Informationen aus den Aufschlüssen heraus.

In den dichten, fluviatilen Sandsteinen des Oberkarbons ist ein besseres Verständnis der Variationen in der Reservoirqualität essenziell für die Entwicklung erfolgreicher Explorationsstrategien.

Zwei Aufschlüsse dieser Oberkarbon Sandsteine (Westphal D), die als Analog zu den produzierenden Lagerstätten im Norden des Niedersächsischen Beckens angesehen werden können, sind im Südteil des Beckens untersucht worden. Der Woitzel Steinbruch, unteres Westphal D, liegt in einer Entfernung von etwa 15 km und stratigraphisch rund 600 m unter dem Hüggel Aufschluss, oberes Westphal D.

Bessere Reservoirqualitäten der oberen Westphal D Sandsteine am Hüggel im Vergleich zu den unteren Westphal D Sandsteinen im Steinbruch Woitzel sind mit dem Auftreten authigener Kaolinite in den Hüggel Proben assoziiert. Im Gegensatz dazu führt eine intensive Illitisierung der Woitzel Sandsteine zu einer deutlichen Reduktion der Reservoirqualität. Für Proben mit einer Permeabilität von mehr als 0.01 mD besteht eine positive Korrelation mit der Korngröße. Unterschiede in der Permeabilität von zwei Größenordnungen (0.001 – 0.1 mD) über eine laterale Entfernung von 150 m werden außerdem mit der sedimentären Geometrie der fluviatilen Sandsteine in Verbindung gebracht.

Daten eines dritten Aufschlusses (Piesberg) und von zwei Gasfeldern (Bohrung A und B), die alle das Westphal C/D umfassen, sind zusätzlich integriert worden, um anhand der petrophysikalischen Ergebnisse und paragenetischen Zusammensetzung in Bezug zur Versenkungsgeschichte Rückschlüsse auf beckenübergreifende Kontrollfaktoren auf die Reservoirqualität zu ziehen.

Die Woitzel und Hüggel Sandsteine sowie die Proben aus den Bohrungen zeigen vergleich bare Matrixpermeabilitäten, aber Unterschiede in den Matrixporositäten. Damit einher gehen Unterschiede während der Versenkungsgeschichte. Eine verstärkte Hebung von Feld B im Zuge der Beckeninversion erklärt das reduzierte Bildungspotenzial von authigenen Quartzzementen im Vergleich zu Feld A und ist der Hauptkontrollfaktor für die höheren

Contents

Acknowledgements i

Abstract iii

Kurzfassung vii

Declaration of Originality xv

1 Introduction 1

1.1 Objective 1

1.2 Aims 2

1.3 Correlating Upper Carboniferous and Zechstein Ca2 reservoir properties 3

1.4 Overview of the thesis 5

1.4.1 Porosity and permeability variations of Upper Carboniferous outcrops (chapter 2) 5

1.4.2 Reservoir quality development of Upper Carboniferous outcrop and reservoir rocks (chapter 3) 6

1.4.3 Correlation of Zechstein Ca2 deposits and assessment of a suitable outcrop analog (chapter 4) 6

1.4.4 Incorporation of terrestrial laser scanning to compare Zechstein Ca2 outcrop and reservoir fracture statistics (chapter 5) 7

1.4.5 Control of the present-day stress field on Zechstein Ca2 fractures (chapter 6) 7

1.5 Parts of this thesis which have been published 7

2 Porosity and permeability variations in a tight gas sandstone reservoir analog, Westphalian D, Lower Saxony Basin, NW Germany: Influence of depositional setting and diagenesis 11

2.1 Abstract 11

2.2 Introduction 12

2.3 Geological setting 14

2.3.1 Study area 17

2.4 Materials and methods 18

2.5 Results 19

2.5.1 The lower Westphalian D succession in the Woitzel quarry 19

2.5.2 The upper Westphalian D succession at Hüggel 24

2.5.3 Petrography 25

2.5.4 Petrophysics 33

2.6 Discussion 38
2.6.1 Paragenetic sequence 40
2.6.2 Factors controlling reservoir quality 45
2.7 Conclusions 50

3 Outcrop to well – Integrating the reservoir quality evolution of Upper Carboniferous tight gas sandstone analogs and reservoirs, NW Germany 51
3.1 Abstract 51
3.2 Introduction 52
3.3 Geological setting 53
3.4 Materials and methods 55
3.5 Results 58
3.5.1 Lithology 58
3.5.2 Petrography 61
3.5.3 Thermal exposure 64
3.5.4 Petrophysics 65
3.6 Discussion 67
3.6.1 Variations in paragenesis 67
3.6.2 Reservoir quality control on a basin-wide scale 71
3.6.3 Reservoir quality control on an intra-field scale 73
3.6.4 Implications for exploration strategies 74
3.7 Conclusions 75

4 Correlating three Zechstein Ca2 carbonate outcrops across the Eichsfeld-Altmark Swell – Facies, reservoir properties, and outcrop analog potential 77
4.1 Abstract 77
4.2 Introduction 78
4.3 Geological setting 79
4.4 Materials and methods 81
4.5 Results 82
4.5.1 Ca2 carbonates of Quarry 1 82
4.5.2 Ca2 carbonates of Quarry 2 83
4.5.3 Ca2 carbonates of Quarry 3 85
4.5.4 Petrophysical properties 86
4.6 Discussion 87
4.6.1 Gross depositional environment interpretation 87
4.6.2 Correlation across the Eichsfeld-Altmark Swell 89
4.6.3 GDE vs reservoir characteristics 90
4.7 Conclusions 93

5 Comparing fracture statistics from outcrop and reservoir data using conventional manual and t-LiDAR derived scanlines in Ca2 carbonates from the Southern Permian Basin, Germany 95
5.1 Abstract 95
5.2 Introduction 96
5.3 Geological setting 98
5.4 Materials and methods 100
5.4.1 Fracture detection using t-LiDAR scanning 102
5.5 Results 104
5.5.1 Ca2 carbonates 104
5.5.2 Fractures 108
5.6 Discussion 118
5.6.1 Validation of outcrop and reservoir comparability 118
5.6.2 Fracture history 120
5.6.3 Validation of t-LiDAR 122
5.6.4 Added values for reservoir modeling 125
5.7 Conclusions 129

6 Present-day stress control on fluid migration pathways: Case study fractured Zechstein carbonates, NW-Germany 131
6.1 Abstract 131
6.2 Introduction 131
6.3 Geological setting and tectonic regime 133
6.4 Materials and methods 136
6.4.1 Dilation tendency and slip tendency 136
6.5 Results 139
6.5.1 Subsurface fracture data 139
6.5.2 Dilation- and slip tendency results for scenario 1 140
6.5.3 Dilation- and slip tendency results for scenario 2 141
6.6 Discussion 142
6.6.1 Controlling factors on Td and Ts 142
6.6.2 Implications for reservoir modeling 144
6.7 Conclusions 147

7 Conclusions and Outlook 149
7.1 Conclusions 149
7.2 Outlook 151

References 153

Appendix 179

Declaration of Originality

Chapter 2: As first author during her time as PhD student, Ivy Becker extended and re-evaluated the dataset prepared by her in her graduate project. She wrote the entire article with inputs from Patrick Wüstefeld, Bastian Koehrer, Marita Felder, and Christoph Hilgers. The discussions with all co-authors contributed to this manuscript.

Chapter 3: As first author, Ivy Becker extended, combined, and re-evaluated the dataset prepared for chapter 2. Assistance during point-count analyses and evaluation by co-author Benjamin Busch is highly appreciated. She wrote the entire article with inputs from Benjamin Busch, Bastian Koehrer, Dirk Adelmann, and Christoph Hilgers. The discussions with all co-authors contributed to this manuscript.

Chapter 4: As first author, Ivy Becker performed the data analysis and evaluation. A subset of the plug samples were petrophysically measured by Marius Waldvogel and Simon Freitag for their graduate projects and quality-controlled by the author. She wrote the entire article with input from Bastian Koehrer and Christoph Hilgers. The discussions with both co-authors contributed to this manuscript.

Chapter 5: As first author, Ivy Becker performed the data acquisition, fracture analysis and evaluation. Parts of the manual fracture measurements and fracture detection of an aerial photograph were performed by Marius Waldvogel for his graduate project and quality-controlled by the author. She wrote the entire article with inputs from Bastian Koehrer, Marius Waldvogel, Wolfgang Jelinek, and Christoph Hilgers. The discussions with all co-authors contributed to this manuscript.

Chapter 6: As first author, Ivy Becker performed all data analysis and evaluation. She wrote the entire article with input from Birgit Müller, Bastian Koehrer, Wolfgang Jelinek, and Christoph Hilgers. The discussions with all co-authors contributed to this manuscript.

1 Introduction

1.1 Objective

In NW Europe, large hydrocarbon reserves are stored in tight sandstone and carbonate formations below and within the Late Permian (Zechstein) evaporite sequences. Upper Carboniferous fluvial sandstone- and Zechstein Ca2 (Stassfurt, second cycle) carbonate reservoirs have been successfully produced for the last decades, but especially the first are known to still bear great exploration potential (e.g., Besly, 1998; Jones and Glover, 2005; Karnin et al., 1992; Kombrink et al., 2010; Leeder and Hardman, 1990; Strohmenger et al., 1996; Strohmenger and Strauss, 1996). First, their stratigraphic position below the thick Zechstein evaporite sequences, reaching thicknesses of several hundreds of meters (e.g., Peryt et al., 2010), impedes exploration attempts in those formations as the quality of seismic images is reduced (e.g., Besly, 1998). Secondly, a complex interaction of tight matrix reservoir characteristics, controlled by the depositional setting and diagenetic history, and structural inventory, i.e. faulting and fracturing, complicates the attempts to correctly predict reservoir quality of the target formation.

Regarding the Upper Carboniferous fluvial tight gas-producing sandstone reservoirs, lateral variations in reservoir characteristics are already initiated by the primary depositional setting with varying channel bar geometries or facies variations (cf. Kombrink et al., 2010; Miall, 1988; Pranter et al., 2007). Thus, an improved understanding of those lateral variations, from internal grain-size trends (e.g., Miall, 1988; Pranter et al., 2007) to inter-bar connectivity is key for future well planning. In addition, a detailed study of the diagenetic evolution in relation to the primary depositional facies and the burial history is needed to explain variations in the matrix reservoir characteristics within one unit and of different reservoirs. Generally, a strong diagenetic overprint is reported for the Upper Carboniferous fluvial sandstones (Besly, 1998; Lüders et al., 2010). In those tight sandstones with a low matrix permeability, the presence of faults and fractures can provide additional fluid migration pathways, which is why many exploration attempts were focused on inverted block structures (e.g., Hollmann et al., 1997; Kombrink et al., 2010; Lüders et al., 2010). However, it has been shown that those fault-related targets may be associated to a thermal overprint of the formation, that can be related to the circulation of hydrothermal fluids along faults (Wüstefeld et al., 2017b). The occurrence of such thermal anomalies in the area of the northern German Upper Carboniferous fields has been studied in detail to better understand their formation (e.g., Cameron and Ziegler, 1997; Lüders et al., 2012; Petmecky et al., 1999; Senglaub et al., 2005a; Senglaub et al., 2005b; Teichmüller, 1986).

With regard to the Zechstein Ca2 gas-producing carbonate reservoirs, many studies have been focused on their facies distribution along the southern margin of the Southern Permian Basin proposing a general division into basin-, slope-, and platform facies with various sub-types (e.g., Geluk, 2000; Reijers, 2012; Strohmenger et al., 1996; Strohmenger et al., 1998; Strohmenger and Strauss, 1996). The diagenetic evolution, and especially the different stages of dolomitization and calcitization, strongly affect the porosity characteristics of those carbonates, and their paragenetic sequences have been widely published (e.g., Biehl et al., 2016; Clark, 1980, 1986; Reijers, 2012; Strohmenger et al., 1998). Those diagenetic alterations were shown to be traceable over the entire Southern Permian Basin (Clark, 1980), but differences occur between the depositional facies types (Strohmenger et al., 1998). However, most carbonate reservoirs occur as naturally fractured reservoirs (Lamarche et al., 2012), and thus, fractures control the flow in those otherwise tight rocks. Understanding the fracture network geometries significantly improves reservoir characterization and well planning (e.g., Gale et al., 2005; Gillespie et al., 2011; Lamarche et al., 2012; Lavenu et al., 2013; Narr, 1996; Peacock et al., 2003; Wennberg et al., 2006; Wilson et al., 2011). As fractures are generally below the resolution limit of seismic images (Gillespie et al., 2011; Lamarche et al., 2012; Seers and Hodgetts, 2014; Wennberg et al., 2006), and vertical wells only correctly represent certain fracture orientations (Narr, 1996; Terzaghi, 1965), understanding the structural inventory is challenging especially during early field development and further studies are needed.

Both, the need to gain knowledge about lateral variations in fluvial sandstone deposits and the necessity of additional fracture data point to incorporating the use of outcrop analogs. Such outcrops must resemble the lithology and depositional environment to serve as analogs for the reservoir rocks, and further, they must feature similar diagenetic alterations to guarantee the comparability of fracture mechanics (Gale et al., 2005; Laubach et al., 2009; Lavenu et al., 2013; Wennberg et al., 2006). Scarce surface exposures with potential analogous characteristics for the studied rocks are present (cf. Besly, 1998; Kombrink et al., 2010; Peryt et al., 2010), but there are rare outcrops of Upper Carboniferous fluvial sandstones, located on the SW border of the Lower Saxony Basin, and of Zechstein Ca2 carbonates along the SW margin of the Harz mountains, that have been shown to bear great potential to be incorporated as analogs.

1.2 Aims

The aim of this study is to better understand the controlling factors on reservoir quality of the Upper Carboniferous fluvial sandstone- and Zechstein Ca2 carbonate reservoirs in NW Germany. Therefore, the suitability of outcrop analogs for both formations is validated, and their characteristics are studied in detail. Petrophysical and petrographic data are linked to the burial histories and structural characteristics to explain the reservoir quality development of the studied rocks. In a next step, reservoir data of the associated formations is incorporated and similarities and differences are discussed. Furthermore, the influence of the present-day stress field on the subsurface reservoir fractures in contrast to outcrop fractures is addressed. In this work, the focus is on the following aspects:

Relating the porosity and permeability variations of Upper Carboniferous fluvial sandstone outcrop analogs to their depositional environment and diagenetic evolution to explain reservoir heterogeneities from the scale of a single sand bar unit to a large stratigraphic scale.

- Linking Upper Carboniferous outcrop and reservoir data to show different paths of reservoir quality development by incorporating diagenetic and structural aspects.
- Correlating Zechstein Ca2 fracture pattern of outcrop and reservoir data to discuss the transferability of fracture data obtained from outcrops to reservoir modeling.
- Validating the application of terrestrial laser scanning (t-LiDAR) for fracture data generation to improve the collection of large amounts of data for statistical analyses.
- Modeling the influence of the present-day stress field on reservoir fractures to show the impact on reservoir flow characteristics, and thus, ultimately on the calculation of reservoir production performance.

1.3 Correlating Upper Carboniferous and Zechstein Ca2 reservoir properties

The Upper Carboniferous fluvial sandstones and Zechstein Ca2 carbonates are all studied petrographically to derive diagenetic mineral alterations, and petrophysically, deriving their porosity and permeability characteristics with the help of He-pycnometry and an isostatic flow cell.

With regard to the Upper Carboniferous, two outcrops (40 thin sections, 48 porosity and permeability measurements) are investigated. Data of a third outcrop (51 thin sections, 36 porosity and permeability measurements) presented by Wüstefeld et al. 2017a and b is incorporated to compare the results. In addition, 94 thin section of two Upper Carboniferous wells of two fields are analyzed to be linked to the provided petrophysical data (86 porosity and permeability measurements). As no oriented fracture data is available from the reservoirs, this part of the study mainly concentrates on the matrix porosity and permeability development in relation to the diagenetic history. In one of the studied outcrops, 441 fracture orientations are additionally measured.

With regard to the Zechstein Ca2, porosity and permeability values of 75 samples of three outcrops are measured and qualitative petrographic analyses are performed on 51 thin sections. As fracture data from resistivity image log data of a horizontal well is provided, containing a total of 1050 open fractures and 98 cemented fractures, and only three thin sections and 15 porosity and permeability values of a reference well in the same field are available, here, the focus of this work is on the structural analysis. Therefore, a total of 1118 open fractures were manually measured in the outcrops.

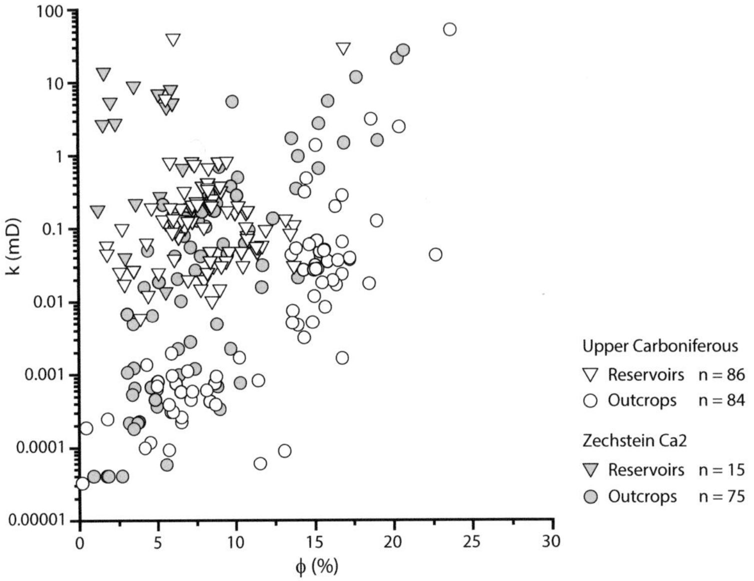

Figure 1.1. Porosity-permeability cross-plot of the Upper Carboniferous outcrop (n=84, white circles, chapter 2 and 3) and subsurface (n=86, white triangles, chapter 3) sandstone samples compared to the Zechstein Ca2 outcrop (n=75, gray circles, chapter 4) and subsurface (n=15, gray triangles, chapter 4) carbonate samples of this study. Here, all values are given for 2 MPa confining pressure.

The Upper Carboniferous sandstones and the Zechstein Ca2 carbonates each show a strong diagenetic overprint that resulted in the tight matrix porosity and permeability characteristics of both formations (Fig. 1.1). In both settings, the occurrence of fractures results in reservoir quality enhancement by providing potential hydrocarbon migration pathways. A comparison of the fracture data of the Upper Carboniferous lower Westphalian D outcrop and the Zechstein Ca2 outcrops shows a main NNE – SSW to NE – SW oriented fracture set and less abundant approximately W – E oriented fracture sets in both settings (Fig. 1.2). NW – SE oriented fractures are more pronounced in the Zechstein Ca2 data set. However, this comparison of fracture orientations of Upper

Carboniferous sandstones at the southern margin of the Lower Saxony Basin and of Zechstein Ca2 carbonates at the SW margin of the Harz mountains indicate a regional fracture pattern.

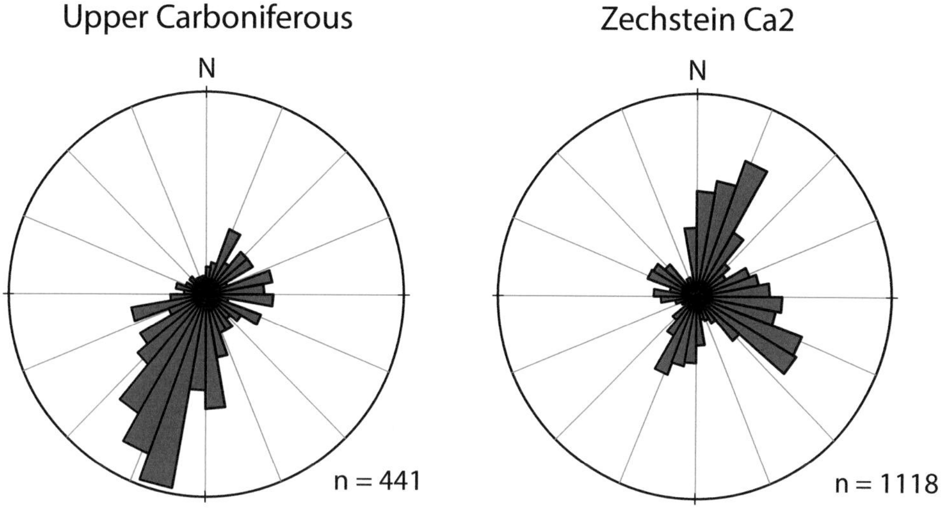

Figure 1.2. Manually measured fracture strike orientations of an Upper Carboniferous outcrop (chapter 2 and 3, Woitzel quarry) on the southern margin of the Lower Saxony Basin (n=441) compared to manually measured strike orientations of the three studied Zechstein Ca2 outcrops (chapter 4) along the SW margin of the Harz mountains (n=1118) plotted as rose diagrams. Results indicate similar main NE – SW fracture orientations.

1.4 Overview of the thesis

1.4.1 Porosity and permeability variations of Upper Carboniferous outcrops (chapter 2)

In this chapter, porosity and permeability characteristics of lower Westphalian D sandstones are compared to upper Westphalian D sandstones. The petrophysical data are linked to petrographic data obtained from point count analyses. A reduction in permeability by one order of magnitude depending on the stratigraphic position of the two locations is related to the present authigenic clay type. Samples from the upper Westphalian D, containing kaolinite, have higher permeabilities than samples from the lower Westphalian D, containing illite. This occurrence of different authigenic clay types, i.e. illite and kaolinite, is further associated with differences in depositional conditions that resulted in variations in the original feldspar content between the lower and upper Westphalian D.

Higher original feldspar contents are present in the lower Westphalian D sandstones. Within one location, lateral variations in permeability by two orders of magnitude are shown over a lateral distance of 150 m in a single sandstone unit. In addition, a positive correlation between permeability and grain size is shown within one location for samples with permeability values above 0.01 mD.

1.4.2 Reservoir quality development of Upper Carboniferous outcrop and reservoir rocks (chapter 3)

This chapter integrates three more data sets (one additional outcrop- and two reservoir data sets, all Westphalian C/D) to the current lower and upper Westphalian D outcrop data sets. Authigenic cementation processes, reducing permeability and porosity, such as the formation of illite and quartz cement are temperature-dependent (e.g., Bjørlykke and Egeberg, 1993; Busch et al., 2018; Makowitz et al., 2006; Walderhaug, 1994, 2000; Worden and Burley, 2003). Regarding the formation of quartz cements, it has been shown to be sensitive to the time spent at elevated temperatures high enough for its formation (~75°C, Walderhaug, 1994). In addition, grain size, available surface area, and the attainment of euhedral crystal facets also control quartz cementation (e.g., Busch et al., 2018; Lander et al., 2008; Makowitz et al., 2006; Walderhaug, 1994, 2000). Therefore, the thermal exposure of each study area was established and it is shown that the amount of authigenic quartz cements depend on those thermal differences. Differences in the porosity evolution are mainly controlled by the amount of quartz cement formation and uplift to the surface results in additional matrix porosity due to the dissolution of carbonate cements.

1.4.3 Correlation of Zechstein Ca2 deposits and assessment of a suitable outcrop analog (chapter 4)

This chapter introduces three Zechstein Ca2 outcrops along the SW margin of the Harz mountains. Exposed rock characteristics are linked to a depositional facies type related to the position on a carbonate platform to slope environment of the southern margin of the Southern Permian Basin. Based on that facies classification, the three outcrops are correlated across the 'Eichsfeld-Altmark Swell', a NNE – SSW striking paleo-high on which the carbonates were deposited. Petrophysical data are further linked to the interpreted facies types and platform margin facies show best reservoir characteristics. Due to their variations in reservoir quality, a potential analog setting is indicated for each potential reservoir facies.

1.4.4 Incorporation of terrestrial laser scanning to compare Zechstein Ca2 outcrop and reservoir fracture statistics (chapter 5)

In this chapter, a Zechstein Ca2 carbonate quarry at the SW margin of the Harz mountains is shown to be a suitable outcrop analog for a gas-producing Ca2 reservoir in NW Germany. They represent a similar lithology of dolo-mud- to dolo-wackestones, a similar depositional environment on the slope of the former southern margin of the Southern Permian Basin, and comparable diagenetic alterations. Based on these similarities, fracture characteristics of both, outcrop and reservoir, are compared revealing similar main orientations (NNE – SSW to NE – SW) that suggest a preferential W – E well orientation. Variations in mean main strike orientations between the outcrop and reservoir are discussed to be related to near-field stress perturbations around the faults bounding the reservoir or the position of the outcrop on a NNE – SSW striking paleo-high. Terrestrial laser scanning (t-LiDAR) data is incorporated using a semi-automated workflow firstly applied to carbonate rocks. Fracture data are successfully validated by comparison with conventional compass measurements.

1.4.5 Control of the present-day stress field on Zechstein Ca2 fractures (chapter 6)

This chapter adds the analysis of the actual hydraulic conductivity potential of the reservoir fractures to the analysis of the fracture network geometries presented in the previous chapter. The influence of the present-day stress field acting on the subsurface fractures is modeled as slip- and dilation tendency analyses. Such analyses allow quantification of the critical stress state of a fracture surface within a stress field (e.g., Collettini and Trippetta, 2007; Ferrill et al., 1999; Lisle and Srivastava, 2004; Morris et al., 1996; Morris and Ferrill, 2009; Neves et al., 2009). As no stress data are available for the reservoir in the focus of this study, two different scenarios are calculated based on stress data of a nearby field. Results indicate a strong reduction of the actual fracture permeability for the calculated control of both defined stress fields on the opening potential of the reservoir fractures.

1.5 Parts of this thesis which have been published

Articles:

Becker, I., Koehrer, B., Waldvogel, M., Jelinek, W., Hilgers, C., 2018, Comparing fracture statistics from outcrop and reservoir data using conventional manual and t-LiDAR derived scanlines in Ca2 carbonates from the Southern Permian Basin, Germany, Marine and Petroleum Geology 95, 228-245.

Becker, I., Wüstefeld, P., Koehrer, B., Felder, M. and Hilgers, C., 2017, Porosity and permeability variations in a tight gas sandstone reservoir analogue, Westphalian D, Lower Saxony Basin, NW Germany: Influence of depositional setting and diagenesis, Journal of Petroleum Geology, 40(4), 363-389.

Submitted manuscripts:

Becker, I., Busch, B., Koehrer, B., Adelmann, D., Hilgers, C., subm., Outcrop to well - Integrating the reservoir quality evolution of Upper Carboniferous tight gas sandstone analogues and reservoirs, NW Germany.

Becker, I., Müller, B., Koehrer, B., Jelinek, W., Hilgers, C., subm., Present-day stress control on fluid migration pathways: Case study Zechstein fractured carbonates, NW- Germany.

Becker, I., Koehrer, B., Hilgers, C., subm., Correlating three Zechstein Ca2 carbonate outcrops across the Eichsfeld-Altmark Swell – Facies, reservoir properties, and outcrop analog potential.

Additional submitted manuscripts:

Busch, B., Becker, I., Koehrer, B., Adelmann, D., Hilgers, C., subm., Porosity evolution of two Upper Carboniferous tight-gas-fluvial sandstone reservoirs: Impact of fractures and total cement volumes on reservoir quality.

Conference contributions:

Becker, I., Wüstefeld, P., Koehrer, B., Busch, B., and Hilgers, C., 2018, Developing an exploration strategy for Upper Carboniferous tight gas sandstone reservoirs, at EAGE conference and exhibition 2018 in Copenhagen.

Becker, I., Koehrer, B., Jelinek, W., and Hilgers, C., 2018, Outcrop to well – Transferability of Zechstein Ca2 Carbonates from the Eichsfeld-Altmark Swell to the North German Subsurface, at DGMK Frühjahrstagung 2018 in Celle.

Becker, I., Wüstefeld, P., Koehrer, B., Jelinek, W., and Hilgers, C., 2017, Comparison of open fractures in Ca2 carbonates from outcrop and well, using laser scans, image logs, and 3D stress, at EAGE Carbonates Workshop 2017 in Potsdam.

Becker, I., Wüstefeld, P., Koehrer, B., Jelinek, W., and Hilgers, C., 2017, Digitally derived fracture network relationships in fractured Zechstein carbonate reservoirs from surface and well data, at EAGE conference and exhibition 2017 in Paris.

Becker, I., Wüstefeld, P., Koehrer, B., Jelinek, W., and Hilgers, C., 2016, Fracture network relationships in Ca2 carbonates from outcrop and well data in the Southern Permian Basin, Germany, at DGMK Frühjahrstagung 2017 in Celle.

Becker, I., Wüstefeld, P., Hilse, U., Manss, Y., Koehrer, B., and Hilgers, C., 2016, Reservoir heterogeneity of tight gas sandstones from observations of outcrop and subsurface data in the Lower Saxony Basin, Germany, at DGMK Frühjahrstagung 2016 in Celle.

Becker, I., Wüstefeld, P., Hilgers, C., and Koehrer, B., 2015, Sealing and leaching processes around normal faults in Upper Carboniferous tight gas sandstones from outcrop analog studies, Geotectonic Research 97, Special Issue DRT 2015 Abstracts, p. 123.

Becker, I., Wüstefeld, P., Hilgers, C., and Koehrer, B., 2015, Reservoir heterogeneity in tight gas sandstones – Comparing outcrop analog studies with subsurface data, at Student Technical Conference STC 2015 in Wietze.

Becker, I., Wüstefeld, P., Hilgers, C., and Koehrer, B., 2015, Reservoir heterogeneity in Upper Carboniferous tight gas sandstone reservoirs, at Aachener Geotag 2015.

2 Porosity and permeability variations in a tight gas sandstone reservoir analog, Westphalian D, Lower Saxony Basin, NW Germany: Influence of depositional setting and diagenesis

2.1 Abstract

Understanding the spatial variability of reservoir properties in tight gas sandstones can significantly reduce uncertainties in reservoir characterization. This study focuses on two outcrops of Upper Carboniferous (Westphalian D) sandstones in the southern part of the Lower Saxony Basin, NW Germany, which are analogs for tight, gas-producing reservoir rocks in the north of the basin. Large-scale differences in reservoir quality occur between the two outcrops which are separated by a distance of approximately 15 km and by 600 m of stratigraphy; smaller-scale heterogeneity in the form of channel geometries occurs within individual fluvial cycles, as evidenced in a 40 m high and a 150 m long porosity-permeability profile.

In the Woitzel quarry, lower Westphalian D strata consist of fining-upwards cycles of fluvial sandstones with intercalated floodplain sediments and coal seams. In abandoned quarries at the Hüggel location, Upper Westphalian D strata are composed of red and gray colored sandstones which suggest deposition in a fluvial – alluvial plain environment under increasingly arid conditions.

Reservoir quality is higher in the upper Westphalian D sandstones at Hüggel (permeability: up to 1 mD; mean porosity: 19%) than in the lower Westphalian D sandstones at Woitzel (permeability: <0.1 mD; mean porosity: 15%). Due to the high degree of compaction of the upper and lower Westphalian D deposits (intergranular volumes of 21.9% and 19.4% respectively) and the high proportions of pseudomatrix, porosity in these sandstones is mainly secondary with up to 80% microporosity. Reservoir quality is in general improved by the presence of kaolinite in the upper Westphalian D sandstones. In contrast, intense illitization has degraded the reservoir quality of the lower Westphalian D sandstones.

A positive correlation between permeability and grain size is exhibited by the kaolinite-bearing upper Westphalian D sandstones. No such correlation is evident in the illite-bearing lower Westphalian D sandstones, unless samples with permeabilities of <0.01 mD are excluded from the correlation. However, such a low permeability cut-off is not related to sorting, grain size or authigenic quartz cementation.

At a small scale, permeability varies laterally by two orders of magnitude from 0.001 mD to 0.1 mD in a single, approximately 3 m thick tight sandstone bed over a distance of 150 m. Repeated, centimeter-scaled fining-upwards grading and bed-internal erosion contribute to lateral variations in reservoir flow characteristics.

Integrated studies of the depositional environment, diagenesis and sedimentary geometries of the deposited units are a key to the understanding of heterogeneities in tight fluvial sandstones and thus, to the successful development of Upper Carboniferous reservoirs in the Lower Saxony Basin and elsewhere.

2.2 Introduction

Tight, gas-producing sandstone reservoirs are of increasing importance and have been the focus of many recent studies (Dai et al., 2017; Fall et al., 2015; Higgs et al., 2007; Hood and Yurewicz, 2008; House and Shemeta, 2008; Jiang et al., 2015; Laubach and Gale, 2006; Lyu et al., 2017; Olson et al., 2009; Sahoo et al., 2016; Shanley et al., 2004; Stroker et al., 2013; Tobin et al., 2010; Zou et al., 2016). Fluvial sandstones in particular show different scales of heterogeneity due to variations in depositional facies, sedimentary architecture and internal grain-size trends (Miall, 1988; Pranter et al., 2007), resulting in uncertainties in reservoir development.

In general, depositional environments governed by provenance and transport mechanisms control the primary composition of a sandstone and strongly affect the associated reservoir properties (e.g., Bjørlykke, 1998; Dickinson and Suczek, 1979; Estupiñán et al., 2007; Garzanti and Vezzoli, 2003; Haughton et al., 1991; Khalifa and Morad, 2015; Kim et al., 2007; Leeder and Hardman, 1990; McBride, 1963; Morad et al., 2010; Morad et al., 2000; Weltje and von Eynatten, 2004). Reservoir quality is usually relatively high in aeolian and ephemeral fluvial depositional systems (Cowan, 1993; Meadows and Beach, 1993; Olivarius et al., 2015) and in stacked braided channel facies (Moscariello, 2003; Yurewicz et al., 2008), due to relatively large grain sizes (Füchtbauer, 1974; Stroker et al., 2013; Zhang et al., 2015) and/or well-sorted to moderately well-sorted rock fabrics (Bloch, 1991; Olivarius et al., 2015). Diagenetic processes may overprint the primary porosity and permeability trends and may either improve (e.g., via leaching: Besly et al., 1993; Schmidt and McDonald, 1979) or degrade (e.g., via cementation: Cowan, 1993; Dutton and Loucks,

2010) reservoir quality. In fluvial siliciclastic rocks, a high proportion of ductile clasts in the rigid quartz grain matrix leads to a reduction in the primary porosity during mechanical compaction (Bertier et al., 2008; Leeder and Hardman, 1990). Diagenetic deterioration is caused by processes such as e.g. illitization (Cowan, 1993; Gaupp et al., 1993) and quartz overgrowth development (Bjørlykke and Egeberg, 1993; Dutton and Loucks, 2010; Meadows and Beach, 1993; Zhang et al., 2015).

Outcrops can act as analogous for tight, gas-producing reservoir rocks in the subsurface. Outcrop analogs allow lateral variations in rock properties to be studied and can reduce uncertainties in reservoir characterization (cf. McKinley et al., 2011; Miall, 1988; Pranter et al., 2007; Stephen and Dalrymple, 2002; Trendell et al., 2012). McKinley et al. (2011) noted that the permeability in fluvial sandstones can vary by two orders of magnitude at a scale of 1.35x2.85 m, but similar studies at larger scales have not been performed. Outcrops contain information on sedimentary geometries, associated grain size changes, and lateral variability in the diagenetic mineral assemblages, thereby providing information that cannot be obtained from well data alone. Thus, data on lateral changes in sedimentary settings obtained from outcrops are useful for subsurface reservoir modeling.

The NW European Carboniferous Basin hosts degassing hard coal source rocks, seals and siliciclastic reservoir rocks, but major sections remain poorly characterized (Cameron and Ziegler, 1997; Kombrink, 2008). Tight gas reservoir rocks are reported to occur in the southern North Sea (Besly et al., 1993; Leeder and Hardman, 1990; Moscariello, 2003; Ritchie and Pratsides, 1993), in onshore UK and the Netherlands (Kombrink et al., 2007) and in Germany (Henk, 2005; Hollmann et al., 1997; Lüders et al., 2010; Wimmers and Koehrer, 2014) (Fig. 1a). Connected fluvial sandstones are the main reservoir targets. The burial histories of Upper Carboniferous strata in different parts of the basin are characterized by different subsidence rates and thermal overprints, indicating the existence of local palaeo-highs (Bruns et al., 2013; Cameron and Ziegler, 1997; Kombrink, 2008; Petmecky et al., 1999; Senglaub et al., 2005a; Senglaub et al., 2005b; Teichmüller, 1986).

The sedimentology (David, 1990; Jones and Glover, 2005) and fluvial cyclicity (Jones and Glover, 2005) of Westphalian D sandstones at outcrops along the southern margin of the Lower Saxony Basin have been described and these outcrops have been investigated in some detail (Fig. 2.1b). The outcropping sandstones are recognized as potential analogs for the gas fields in NW Germany, the Netherlands, onshore UK and the North Sea (Fig. 2.1a), although reservoir rocks at these fields comprise Westphalian A to D deposits (Jones and Glover, 2005; Kombrink et al., 2010; Moscariello, 2003). However, little information is available on the diagenetic mineral assemblage (Sedat et al., 1992), and the link between petrophysical data and other relevant factors has not been published.

Here, we investigate lateral and vertical variations in porosity and permeability in Upper Carboniferous reservoir analogs and correlate these variations with primary controls (e.g. depositional setting and grain size) and diagenetic factors. We discuss porosity and permeability heterogeneities at different scales, from small-scale variations within a single sandstone unit (10s of meters laterally and several meters vertically) to large-scale stratigraphic trends in lower to upper Westphalian D units (10s of kilometers laterally and 100s of meters vertically). We attempt to evaluate the controlling factors at these different scales.

2.3 Geological setting

The two studied outcrops are located in the Ibbenbüren-Osnabrück hard coal mine district on the SW border of the Lower Saxony Basin (Senglaub et al., 2005a) in NW Germany (Fig. 2.1b). The Upper Carboniferous sandstones were deposited in the NW European Carboniferous Basin, a foredeep NW of the Variscan (Hercynian) orogenic belt (Ziegler, 1989). The Central European Basin System formed as a result of Early Mesozoic rifting and wrench tectonics (Betz et al., 1987; Gast and Gundlach, 2006; Van Wees et al., 2000; Ziegler, 1990). Sub-basins including the Lower Saxony Basin underwent subsidence until the Late Cretaceous followed by subsequent inversion (Betz et al., 1987; Binot et al., 1993; Bruns et al., 2013).

The Upper Carboniferous (Westphalian) coal-bearing sequences represent the principal source rock for the natural gas in the Central European Basin System, which is mainly found in the Permian Rotliegend sandstones and Zechstein carbonates (Glennie, 1986; Kombrink et al., 2010; Leeder and Hardman, 1990; Petmecky et al., 1999). Although successful production of tight gas began more than four decades ago, Upper Carboniferous sandstone reservoirs continue to have considerable exploration potential in NW Germany (Jones and Glover, 2005). Zechstein (late Permian) evaporites represent a regional seal across the entire basin for both, Carboniferous and Permian reservoirs (Kombrink et al., 2010; Petmecky et al., 1999). Upper Carboniferous fluvial sandstones at the two studied outcrops can be regarded as analogs for the subsurface reservoir rocks in gas fields located approximately 50 km to the north (cf. Jones and Glover, 2005; Wimmers and Koehrer, 2014), and are similar in terms of sedimentology, stratigraphy, sedimentary structure and diagenesis.

Fluvial deposition in the Lower Saxony Basin started during the Namurian C and continued through to the Rotliegend (early Permian). The fluvial Westphalian C and lower Westphalian D rocks were deposited on an alluvial plain (Drozdzewski, 2005). Due to hard coal mining over the course of several centuries, the coal seams are well known and some of them can be traced for tens of kilometers (e.g. the Dickenberg seam, Fig. 2.1c)

(Klassen, 1984). The Westphalian units exhibit 4th-order sequences on an outcrop scale (Süss et al., 2000) and reach a total thickness of more than 1500 m in the basin (David, 1990). The cyclic successions are caused by the waxing and waning of the Gondwana ice shield; the magnitude of these successions has been proposed to be <40 m in the northern hemisphere during the Westphalian C and D (Rygel et al., 2008). The large areas inundated by marine transgressions associated with glacio-eustatic sea-level rises suggest that the region was characterized by low relief (Glennie, 1986).

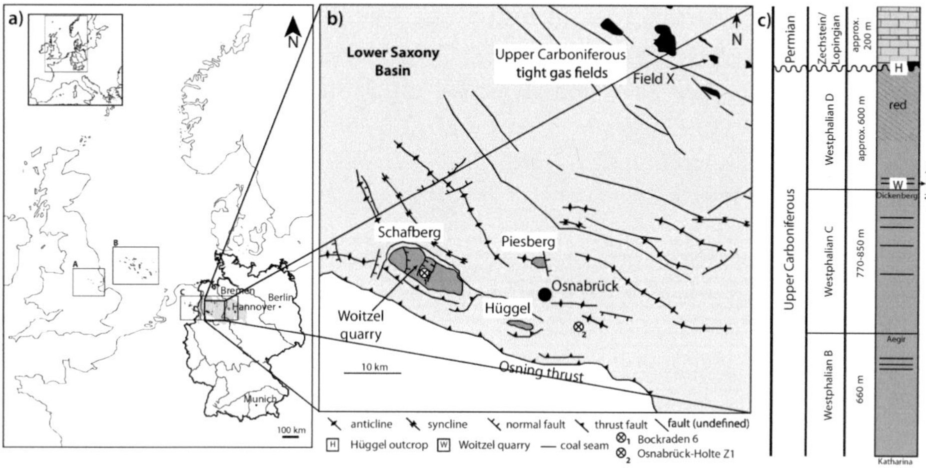

Figure 2.1. a) Location map of fields producing from Upper Carboniferous reservoir units in NW Europe. A: UK onshore; B = North Sea, C = onshore fields in the Netherlands and Germany (modified after Kombrink et al., 2010). Gray shading shows the area of the Lower Saxony Basin in NW Germany. b) Outline map of the southern and central part of the Lower Saxony Basin showing the locations of Upper Carboniferous outcrops at Schafberg, Hüggel and Piesberg (gray), and of tight gas fields producing from Upper Carboniferous reservoir rocks (black). Structures in the southern part of the basin include inversion-related folds, normal faults and the Osning thrust (modified after Drozdzewski, 1985, 1988; Klassen, 1984; Kombrink et al., 2010; Senglaub et al., 2005b). c) The Upper Carboniferous to Permian stratigraphy in the Schafberg area is well known as a result of coal mining (black lines = mineable coal seams including the Alexander seam; David, 1990; Drozdzewski, 1985). The Westphalian D succession is overlain by Permian Zechstein carbonates; the Stephanian corresponds to a hiatus. Sedimentary thicknesses (Drozdzewski, 1985) and the stratigraphic positions of the study areas at the Hüggel (H) and Woitzel (W) quarries, together with the position of the reservoir interval at an actual tight gas field (Field X: location in b), are shown in the stratigraphic column.

In the Lower Saxony Basin, increasingly arid conditions prevailed during the late Westphalian D, as evidenced by the presence of red-colored sand- and siltstones (Besly et al., 1993; David, 1990; Sedat et al., 1992). Floodplain sediments and coal seams were no longer deposited (Besly et al., 1993; David, 1990). The red beds extend to the onshore Dutch fields (Kombrink et al., 2007) and to the southern North Sea and are known as the

"Barren Red Beds" (Besly et al., 1993; Besly and Fielding, 1989; Leeder and Hardman, 1990; Pearce et al., 2005). The red coloration formed either syn-depositionally due to oxidation of a well-drained floodplain environment, or due to later oxidation of originally gray siderite-bearing sandstones during early diagenesis (Besly et al., 1993).

The coarsening-upwards trend from the Westphalian C to the upper Westphalian D is reflected by the percentage of sand, which increases from approximately 60% in the Schafberg area (e.g. 56% in the Westphalian C unit in the Osnabrück-Holte Z1 well, Fig. 2.1b) (Drozdzewski, 2005) to up to 80% (e.g. in the Piesberg quarry, Fig. 2.1b) in the Westphalian D (David, 1987; David, 1990). The sand percentages decrease NW-wards (Jones and Glover, 2005) towards the Dutch fields (Kombrink et al., 2007) (Fig. 2.1a). This trend coincides with general NW-ward Westphalian palaeoflow directions in the Dutch and German onshore sector (Jones and Glover, 2005; Kombrink et al., 2007), for example to the NW in the Woitzel quarry (David, 1990) and to the WNW in the Piesberg quarry (Jones and Glover, 2005) (Fig. 2.1b).

Late Variscan compression (pre-Stephanian) in the Lower Saxony Basin (Betz et al., 1987; Drozdzewski, 1988) was followed by extension, resulting in moderate subsidence and the deposition of the lowermost sediments (Permian) of the Central European Basin System (Bruns et al., 2013; Senglaub et al., 2005a). These sediments consist of the Zechstein (Late Permian) copper shales (a calcareous mudstone which contains copper ore and other metals; Paul, 2006) which rest unconformably on the Westphalian D units along the southern margin of the Lower Saxony Basin (Drozdzewski, 1985). The copper shales are overlain by carbonates (Peryt et al., 2010), and deposition continued until the Late Jurassic (Senglaub et al., 2005a). In the Late Jurassic, the southern Central European Basin System, including the Lower Saxony Basin, began to subside (Bruns et al., 2013; Senglaub et al., 2005a) due to divergent wrench tectonics resulting from rapid crustal extension in the North Sea Rift System (Betz et al., 1987).

In the Lower Saxony Basin, the maximum reconstructed burial temperatures for Upper Carboniferous rocks may be anomalously high, e.g. around 350°C in the Piesberg quarry near Osnabrück (Fig. 2.1b) (Senglaub et al., 2005a), and vary greatly between different outcrops in the Ibbenbüren-Osnabrück coal mine district. The volatile matter contents and vitrinite reflectance values of the Upper Carboniferous coals in the Schafberg complex (Fig. 2.1) indicate that low-volatile bituminous coals with 20% volatile matter are present. Low anthracite coals with 8% volatile matter content are present near the northern rim of the Schafberg complex. Vitrinite reflectance values range from 1.5 to 3% VRm (Klassen, 1984; Teichmüller, 1986). The coals at the Hüggel outcrop are characterized by 16.6% volatile matter, indicating low-volatile bituminous coal (Klassen, 1984; Taylor et al., 1998).

In the Late Cretaceous, basin inversion related to the early Alpine orogeny resulted in the uplift and erosion of up to five kilometers of Lower Saxony Basin fill (Bruns et al., 2013; Kley and Voigt, 2008).

2.3.1 Study area

At the Woitzel quarry location (Fig. 2.2a), Upper Carboniferous (lower Westphalian D) sandstones are exposed in two adjacent quarries which each cover an area of approximately 250 m x 150 m. These quarries are part of the Schafberg complex (Fig. 2.1b), which represents the largest exposure of Upper Carboniferous strata in the Lower Saxony Basin. The second study area, located approximately 15 km to the ESE, consists of a series of abandoned quarries at Hüggel (Fig. 2.2b). The Carboniferous-Permian unconformity is exposed here (Fig. 2.3) (cf. Paul, 1986). The Hüggel complex covers an area of 3.5 x 0.8 km and has been declared a conservation area, so only samples from fallen blocks were collected.

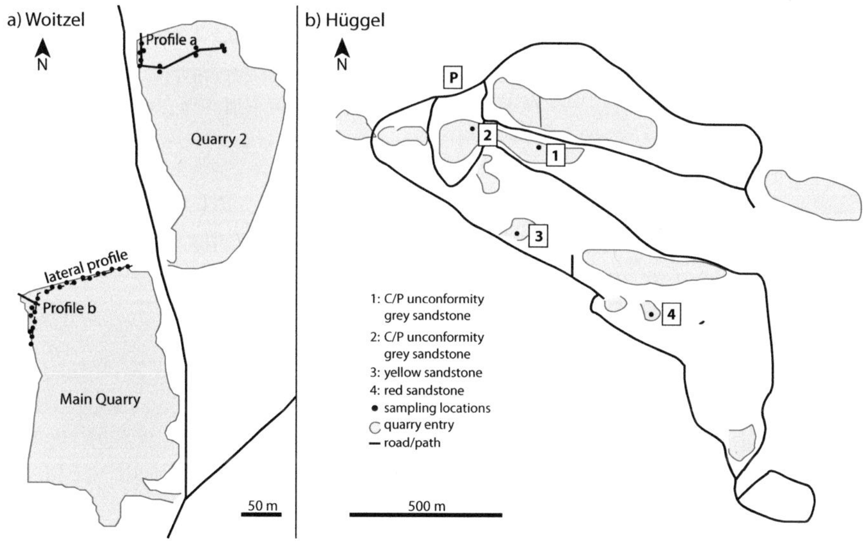

Figure 2.2. Schematic map of the study area in the Lower Saxony Basin: a) The Woitzel quarries ("Main quarry" and "Quarry 2") with the locations of sampling sites and recorded profiles. b) Abandoned quarries in the Hüggel area (gray areas); numbers mark the outcrop locations studied. Regional location in Fig. 2.1b.

The Westphalian D sandstones of the Lower Saxony Basin are in general lithic to sublithic arenites (Sedat et al., 1992) deposited in an alluvial floodplain environment with perennial river systems (David, 1990; Drozdzewski, 2005). Associated depositional facies include channel and bar deposits of different energy levels, and swamp and lake deposits (Jones and Glover, 2005). Stacked (or connected) channel deposits are of economic interest and exhibit unit thicknesses of 10 to 50 m. In general, the Westphalian gas reservoir rocks have 5 - 10% porosity and permeability values in the range of 0.1 to 10 mD (Kombrink et al., 2010; Sedat et al., 1992).

2.4 Materials and methods

In the two Woitzel quarries ("Main Quarry" and "Quarry 2", Fig. 2.2a), three profiles were described (two vertical and one horizontal) and rock sampling and spectral gamma ray measurements were performed. A 30 m high vertical section was recorded in the Main Quarry, and a composite profile spanning 37 m was measured in Quarry 2 (Fig. 2.2a). At the Woitzel quarries, 2.54 cm (1 inch) plugs were drilled parallel to bedding using a water-cooled core drill mounted onto the quarry walls. Twenty-one plugs were collected along a 150 m long horizontal profile in the Main Quarry (Fig. 2.2a), and another 11 plugs were collected along a 37 m high vertical section in Quarry 2.

In the Hüggel area, two vertical profiles, one 10 m high and one 5.5 m high, were logged at sampling locations 1 and 2, respectively (Fig. 2.2b). Fifteen hand specimens representing the differently colored sandstones were collected from four outcrop locations (1-4, Fig. 2.2b). Sample preparation included oven-drying at 60°C for a minimum of 10 days.

Spectral gamma ray values were measured using a hand-held, auto-stabilizing gamma ray spectrometer RS-230 BGO Super-SPEC with a 103 cm^3 bismuth germanate (BGO) detector. The measuring time was three minutes, and the distance between individual measurement positions was <50 cm.

Permeability values (mD) were derived from plugs with diameters of 2.54 cm (1 inch) (33 Woitzel sandstone samples and 15 Hüggel sandstone samples) using single-phase gas flow in an isostatic flow cell. The isostatic flow cell uses helium ($\eta = 1.98*10^{-5}$ Pa*s) as a permeant and water to create confining pressures of up to 35 MPa. The studied plugs were measured at confining pressures of both 10 MPa and 30 MPa to investigate the influence of confining pressure on permeability by determining the associated pressure sensitivity coefficients (cf. David et al., 1994). Helium pressures in the range of 1.5 to 4.5 MPa were adjusted according to sample properties. The porosity (given in %) was measured using helium in a semi-automated gas displacement pycnometer (AccuPyc II 1340 by micromeritics®) and the potential influence of different confining pressures was analyzed using the isostatic flow cell set-up. The statistical error in the results is given as the deviation from the mean value as determined by dividing the standard deviation by the square root of the number of samples.

For petrographic analysis, a total of 29 thin sections from both the horizontal and vertical sections at the Woitzel quarries were examined. Along the horizontal profile, one thin section was prepared for each drilled sandstone plug sample. However, if sandstone samples were taken from a similar stratigraphic height along the vertical profile, only one thin section was prepared. In addition, 11 thin sections from the Hüggel outcrops representing the different colored sandstones at all four sampling locations were prepared. Thin sections were polished and prepared with yellow-dyed, fluorescent, or blue-dyed epoxy-resin. Some samples were stained with Alizarin Red S and potassium ferricyanide to distinguish carbonate cements. Visual analyses were performed using a transmitted-light microscope (Leica DMLP fitted with a Deltapix Invenio 5DII digital camera). All thin sections were point-counted using a Pelcon Automatic Point Counter mounted to a Zeiss Axiolab A1. The distance between each point was dependent on the maximum grain size and to reduce the statistical error, a total of 300 points were counted per thin section. For grain size determination the long axis of a minimum of 100 grains was measured in each thin section using open source software (ImageJ) with grid sizes based on the maximum grain size. The average point count and grain size results are presented as arithmetic means.

To identify differences in mineral composition especially with regard to feldspars and carbonates, a HC1-LM hot-cathode cathodoluminescence microscope with a vacuum of <0.001 mbar, an acceleration voltage of 6 kV and a current density of 0.6 - 0.7 mA was used. The analyses were performed at the Institute of Geology & Palaeontology, RWTH Aachen University. Photographs were taken with a Nikon reflex camera with an automatic exposure meter and the Olympus image software analySIS docu.

Information on sample surface morphology and qualitative elemental analysis was obtained using scanning electron microscopy (SEM) combined with energy dispersive X-ray spectroscopy (EDX). For this purpose, a JEOL JSM-7000F scanning electron microscope equipped with a field emission cathode and an EDAX Pegasus EDX/EBSD system with an acceleration voltage of 15 kV was used at the Central Facility for Electron Microscopy, RWTH Aachen University.

2.5 Results

2.5.1 The lower Westphalian D succession in the Woitzel quarry

Lower Westphalian D fluvial deposits are exposed in the two Woitzel quarries and consist of single- and multi-storey bar sandstones with downstream or lateral accretion surfaces, together with channel sandstones. Two vertical profiles were analyzed (Profiles a and b,

Fig. 2.2a) to describe lateral variations between the quarries. These profiles indicate the presence of fining-upward cycles that pass from clast-supported conglomerates or coarse-grained sandstones to siltstones and grade upwards into coal seams (Figs. 2.3 and 2.4) (see Table 2.1, column 1, for detailed rock characteristics).

Table 2.1. Features of Upper Carboniferous and Zechstein lithologies exposed in the Woitzel (lower Westphalian D) and Hüggel outcrops (upper Westphalian D and Zechstein). Major differences between the sandstone types are underlined.

Lithology	Column 1 Woitzel quarry	Column 2 Hüggel outcrops
	Zechstein	
carbonates	Not exposed	Dark gray mudstone; calcite; >4 m thick
copper shale	Not exposed	Black shale; 0.5-0.6 m thick
	Upper Carboniferous	
coal seams	Black; Alexander seam traceable over both quarries, two minor, laterally limited seams; max. 0.5 m thick	Not exposed
siltstones and shales	Gray; plant fragment and coal flakes; discontinuous parallel bedding; up to 1 m thick	Not exposed
sandstones	Light to medium gray; moderately sorted; medium to coarse-grained; low angle <10°, trough and tabular cross bedded; interval thickness of <15 m Light gray; well to moderately sorted; fine-grained; discontinuous parallel bedding; up to 1.5m thick	Gray and yellow; moderately sorted; fine to coarse-grained; thickness >5 m; in contact with Upper Carboniferous/Zechstein unconformity Red; fine to medium grained; moderately to well sorted; meter-scale cross bedding; bed thickness up to 20 cm, interval thickness >4 m
conglomerates	Gray; intensely weathered; poorly sorted; mostly clast-supported; rounded quartz clasts (Ø = 2 cm), platy mud flakes (l = 5 cm) and organic fragments; sand-sized matrix; bed thickness 0.5 m to >2 m; only local (S-wall Main Quarry, N-wall Quarry 2) and pebble lags	Gray; intensely weathered; poorly sorted; matrix-supported; rounded quartz clasts (Ø = 2 cm); sand-sized matrix; pebble lags (5 cm to 1 m thick) within the sandstone units at outcrops

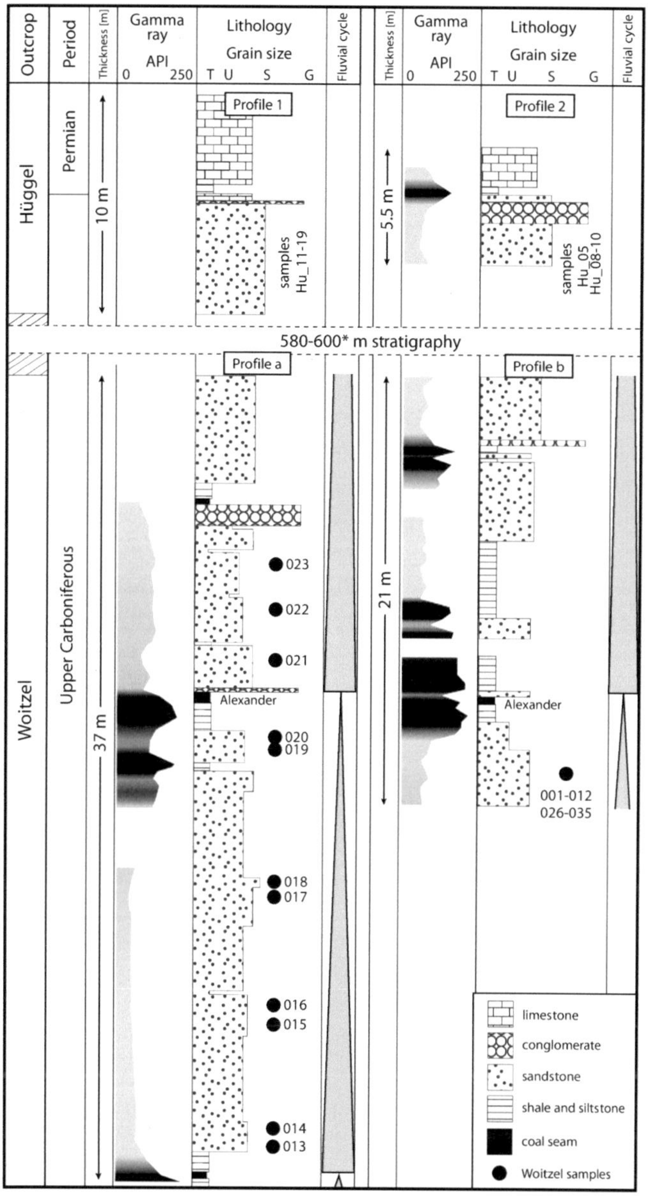

Figure 2.3. Stratigraphic columns including gamma-ray profile (API), lithology and grain size variations, as well as inferred 4th-order fluvial cycles, for the outcrops at Hüggel (Profiles 1 and 2) and Woitzel (Profiles a and b) (see Fig. 2.2 for profile locations). Lateral variations are present at each outcrop. Thus at Woitzel, the middle (Alexander) coal seam is overlain by coarse-grained sandstones and conglomerates in Profile a but by fine-grained, silty material in Profile b. At Hüggel, Profile 1 includes a thin interval of carbonates lying directly above the Carboniferous-Permian unconformity, while Profile 2 includes the copper shale in this position (* stratigraphic thickness after Drozdzewski, 1985). T = clay, U = silt, S = sand, G = gravel.

Clast-supported conglomerates are only rarely present at the bases of a fluvial cycle but do occur locally (e.g. in the south wall of the Main Quarry and the north wall of Quarry 2) and may be more than 2 m thick. Coarse- to medium-grained sandstones represent the principal lithology within a single fluvial cycle and have an interval thickness of up to 15 m (Fig. 2.3). Generally low-angle (<10°) cross bedding is present in coarse- to medium-grained sandstones. Trough and planar-tabular cross bedding are present in the sandstone units below the Alexander coal seam, and in the sandstones above this seam at the top of the measured section along both vertical profiles. Inclined stratification is recorded at both the meter scale and the centimeter scale as evidenced by clay lamination (Fig. 2.4c, e and f). Thin clay-rich intervals show current ripple lamination, and grading from coarse to fine sand sizes can be observed within cm-scale foresets (Fig. 2.4e and f). this internal bedding may parallel the orientation of an entire bar, or may occur at an angle to the larger-scale accretion surfaces, which are more common in the upper parts of the sandstone interval near the Alexander coal seam. The coarse- to medium-grained sandstones grade into fine-grained, well- to moderately sorted gray sandstones with sub-parallel bedding and a total interval thickness of up to 1.5 m, and then grade into overlying siltstones. Thin (0.5 m thick) siltstones and claystones with rootlets underlie the coal seams. In the northern part of the Main Quarry, a thicker siltstone interval (>3 m) overlies the coal and appears to be partly intercalated with thin sandstone units (Fig. 2.3 Profile b, and Fig. 2.4d). Coal seams mark the top of the fluvial cycle. The Alexander coal seam, with a maximum thickness of 0.5 m, is exposed in both Woitzel quarries. In addition, a minor coal seam occurs 25 m below the Alexander seam, and another is exposed 9 m above it in Woitzel Quarry 2 (Fig. 2.3). While the Alexander seam and the seam below it mark the ends of fluvial cycles, the third seam is limited to a length of only 30 m and is only recorded in the northern wall of Quarry 2.

A general fining-upwards trend within each depositional cycle is observed in both quarries. In the sandstones of the Woitzel quarry, the grain size varies more widely within one channel than over the whole sequence (Fig. 2.4e and f). These changes in lithology and grain size within individual fluvial cycles are recorded in the gamma-ray log as lower values (min. 37 API) associated with sandy intervals and higher values (max. 225 API) associated with siltstone and coal intervals with higher clay contents (Fig. 2.3).

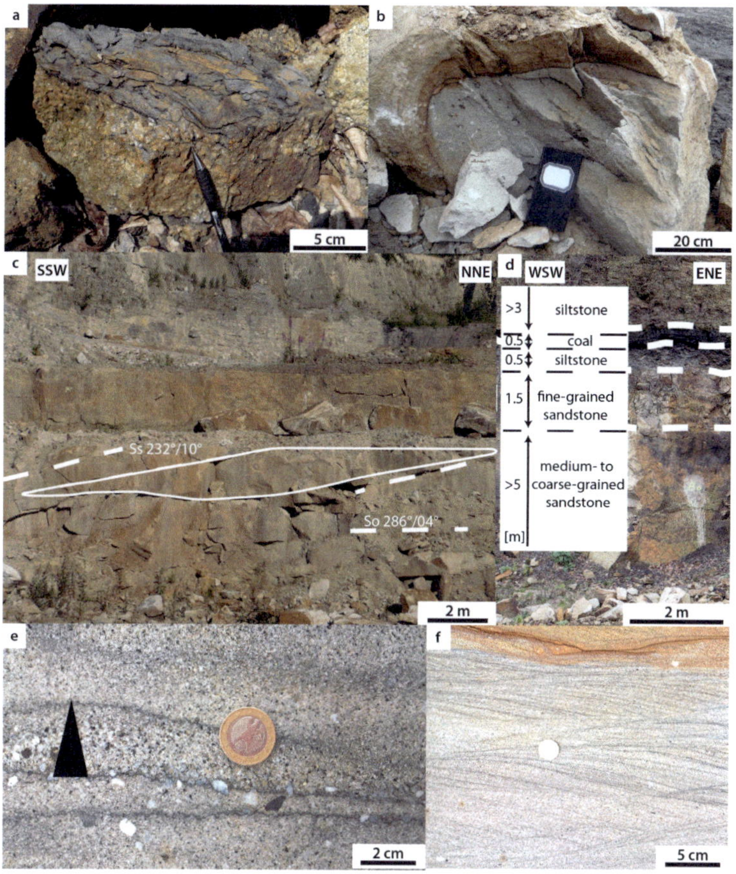

Figure 2.4. Field photographs at the Woitzel quarries. a) Conglomerate with coal fragments. Pen for scale. b) Liesegang bands in a medium-grained sandstone, with a light gray core surrounded by dense dark brown layering. c) Sigmoidal bedforms showing low-angle cross-stratification (Ss: dip direction/dip) dipping towards the SW and bedding orientation (So: dip direction/dip) dipping to the west. d) Typical fluvial succession grading from coarse-grained sandstones to fine-grained sandstones and siltstones, and then to the Alexander coal seam. e) Climbing ripples with centimeter-scale grain-size grading (black arrow) from coarse to finer grains within a point bar deposit. f) Climbing ripples picked out by clay laminae which tend to be oriented at an angle with sigmoidal foresets shown in c).

In the north wall of the Main Quarry at Woitzel, a channel filled with sandy and silty material cuts into the underlying siltstones but not into the underlying Alexander coal seam (Fig. 2.5). The upper channel boundary is defined by the overlying sandstone unit and can be traced on both quarry walls allowing the channel dimensions to be measured and yielding a channel width of approximately 90 m and a maximum depth of 7 m.

Figure 2.5. Field photograph of the northern wall of the Main Quarry at Woitzel (location in Fig. 2a). A fluvial channel is clearly observed in the eastern part of the quarry wall section but becomes obscured to the west due to weathering and vegetation cover (dashed lines). The arrow points to the Alexander coal seam.

2.5.2 The upper Westphalian D succession at Hüggel

In the Hüggel outcrops, the upper Westphalian D sandstones are unconformably overlain by up to 0.6 m of Zechstein (Upper Permian) copper shales and carbonates (Figs. 2.3 and 2.6a, b) (see Table 2.1, column 2, for detailed rock characteristics). The gamma-ray values are high (max. 150 API) for the thin copper shale interval and low (min. 40 API) for the sandstone and carbonate units (Fig. 2.3).

The upper Westphalian D in the Hüggel outcrops is characterized by channel fills deposited in an alluvial plain system. Individual bars could not be distinguished in the exposed succession probably due to weathering. The sandstones can be differentiated based on their color. Those sandstones in direct contact with the Permian unconformity are gray and have a thickness of more than 5 m (Loc. 1 and 2, Fig. 2.2b). Due to intense weathering, no sedimentary structures are preserved in these fine- to coarse-grained sandstones. With increasing distance below the unconformity, the sandstones are yellow due to iron oxide/ -hydroxide precipitation (Loc. 3, Fig. 2.2b and Fig. 2.6c) or red (Loc. 4, Fig. 2.2b). The fine- to medium-grained red sandstones show meter-scale cross bedding with well-defined beds that range from 5 to 20 cm in thickness (Fig. 2.6d). In total, more than 4 m of red sandstones are exposed. The sandstones contain pebbly intervals but lack rootlets or soil features.

Figure 2.6. Field photographs at the abandoned quarries at Hüggel. a, b) Upper Carboniferous fluvial sandstones and conglomerates are unconformably overlain by the Zechstein (Permian) copper shale and carbonates (dashed line = unconformity). c) Intense iron oxide/hydroxide weathering has given the sandstones a yellowish color. d) Reddish sandstones with meter-scale cross-bedding (dashed lines) representing the youngest Carboniferous stratigraphy exposed at this location.

2.5.3 Petrography

The point count data show that both the lower Westphalian D sandstones from Woitzel and the upper Westphalian D sandstones from Hüggel consist of sublithic to lithic arenites with low feldspar contents (Fig. 2.7, Table 2.2). The thin section analyses of samples from Woitzel show that the sandstones are medium- (300 to 400 μm) to coarse- (500 to 600 μm) grained with maximum grain sizes exceeding 2000 μm. The Hüggel sandstones have an average range of fine- to coarse grain sizes from 200 to 600 μm, but the maximum grain sizes are also >2000 μm. The sandstones at both locations are moderately to moderately well sorted. The grains are subangular to subrounded, and larger grains exhibit greater sphericity than smaller ones.

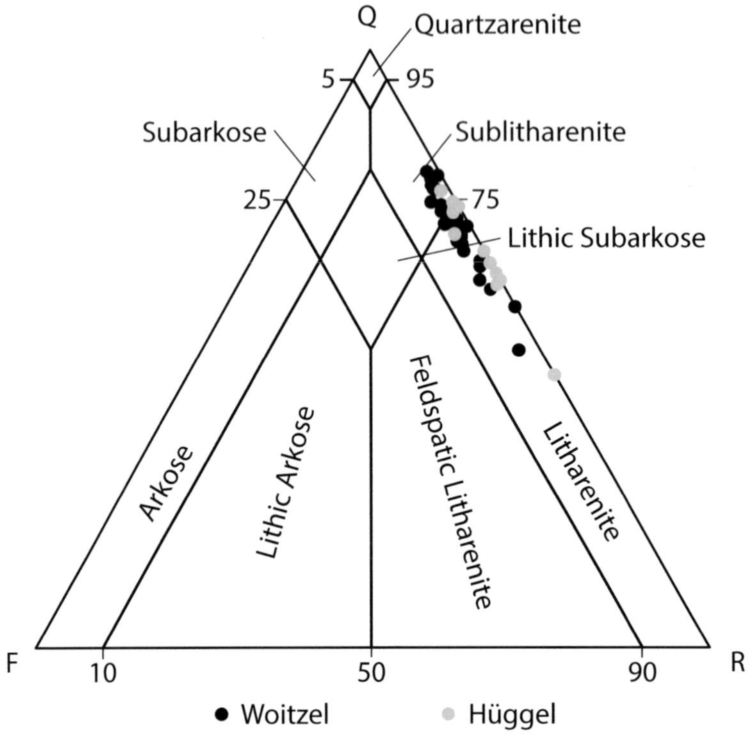

Figure 2.7. Classification of the Westphalian D Woitzel (black) and Hüggel (gray) sandstones on a QFR triangular plot, showing their lithic to sublithic arenite composition (plot after McBride, 1963). In general, the Woitzel sandstones feature slightly higher contents of lithic clasts than the Hüggel sandstones. The plot shows the present-day feldspar content as derived from point counting (Table 2.2).

2.5.3.1 Detrital components

Detrital components in the lower and upper Westphalian D sandstones include both mineral grains and rock fragments (Fig. 2.8). Detrital grains include in descending order of abundance quartz (both mono- and polycrystalline, average 52.7%, max. 61.7%, min. 38%), feldspar (average 1.3%, max. 2.7%, min. 0%), mica (average 0.2%, max. 2.3%, min. 0%), titanite and tourmaline (Table 2.2). The feldspar grains are highly altered and largely replaced by authigenic phases but can be recognized under cathodoluminescence (Fig. 2.8a). Accessory heavy minerals can reach maximum values of 3.7 % and comprise titanite (often situated next to replaced feldspars) and rare tourmaline.

Lithic components include metamorphic rock fragments (average 11.6%, max. 18%, min. 7.3%), shale fragments, coal fragments, iron oxide grains, silicic volcanic rock fragments, siliciclastic rock fragments (feldspar-rich and sandstone fragments), and chert in descending order of abundance. The metamorphic rock fragments include highly deformed quartzite grains, mafic grains that frequently contain quartz veins, and grains that commonly contain illite (Fig. 2.8b and d). The shale fragments show ductile deformation, and their bright yellow interference colors suggest the presence of illite (Fig. 2.8b). The feldspar-rich lithoclasts can be recognized by their multi-mineral composition, in which parts are characterized by strong alteration and illite replacement. In addition, coal fragments are present in samples from the sandstone unit below the Alexander coal seam (Fig. 2.8c).

Figure 2.8. Photomicrographs showing detrital components of the Westphalian D sandstones from Woitzel (Wo) and Hüggel (Hu). a) Monocrystalline quartz grains show straight and undulose extinction. Feldspars are altered and partly replaced by illite along cleavage planes. Cathodoluminescence (CL) of the feldspars (inset) reveals bright red colors suggesting a high Fe3+ content (sample Wo_030). b) Illite-bearing metamorphic rock fragments and shale fragments are common lithoclasts (sample Wo_010). c) Coal fragments occasionally occur in the sandstones below the Alexander coal seam (sample Wo_035). d) Mafic metamorphic rock fragments with quartz vein (sample Wo_035). Qtz = quartz, Fsp = feldspar, Ms = muscovite, (M)RF = (metamorphic) rock fragment, Ill = illite, rp. = replacing, mono = monocrystalline, poly = polycrystalline, und = undulose extinction, xpl = cross-polarized light, ppl = plane polarized light, porosity colored with blue dye.

If individual grains can no longer be distinguished due to a high degree of ductile deformation, the lithoclasts are defined as pseudomatrix (cf. Füchtbauer, 1988; Leeder and Hardman, 1990; Worden and Burley, 2003). Pseudomatrix is abundant (Woitzel: average 5.4%, max. 14%, min. 1%; Hüggel: average 7%, max. 35%, min. 0.7%) and is mostly composed of illite (Fig. 2.9a). In the Woitzel sandstones, kinked mica is often associated with the pseudomatrix highlighting the effects of compaction.

2.5.3.2 Authigenic components

Authigenic iron oxides/-hydroxides are present in samples of the Woitzel sandstones in variable proportions (min. 0%, max. 14.3%), and occur as pore linings or pore space fills and on lithic clasts or meshwork illite. Iron oxides/-hydroxides are also common in sandstone samples from Hüggel in proportions up to 15%. While the hydroxides are amorphous and dispersed around framework grains and within former pore spaces, the iron oxides are granular and are present around or on detrital components (Fig. 2.9b and c).

Illite occurs in pore-filling and fibrous forms (max. 12.3%). In the Woitzel sandstones, authigenic illite fibers occur in feldspars and feldspar-rich rock fragments (Fig. 2.8a), or as the product of the illitization of detrital clay minerals in rock fragments (max. 7.3%) (Fig. 2.9d). Meshwork illite fills secondary pores forming a network of fine needles, and massive flake-like platelets are oriented parallel to each other (Fig. 2.9d-f). In places, illite fibers have grown on tangential illite around detrital quartz grains (Fig. 2.9e). In the Hüggel sandstones, illite occurs as a pore-filling (max. 3.7%) and grain-replacive (max. 2.3%) mineral. The bright interference colors in booklet-shaped kaolinite grains indicate that illite has replaced kaolinite to a minor extent (max. 1.3%) (Fig. 2.9h).

Authigenic quartz forms syntaxial overgrowths (Fig. 2.9b) in sandstone samples from Woitzel (average 3.7%, max. 7.7%, min. 0%) and Hüggel (average 2.6%, max. 5.3%, min. 0.3).

Poikilotopic carbonate cement (max. 15.7%) and kaolinite (max. 11.3%) occur in the Hüggel sandstones but are absent in the Woitzel sandstones. Euhedral crystals of carbonate with dimensions up to 500 µm x 600 µm are present (Fig. 2.10a and b). Back-scattered electron (BSE) imaging with qualitative EDX indicates the presence of two phases of carbonate cement: an outer light gray phase containing Fe, Mg and Ca and an inner darker phase containing Mg and Ca (Fig. 2.10 d,e,f). Staining with Alizarin Red S and potassium ferricyanide distinguished ferrous carbonates, which exhibit blue colors, from dolomitic carbonates, which are unaffected by the staining process (Fig. 2.10c).

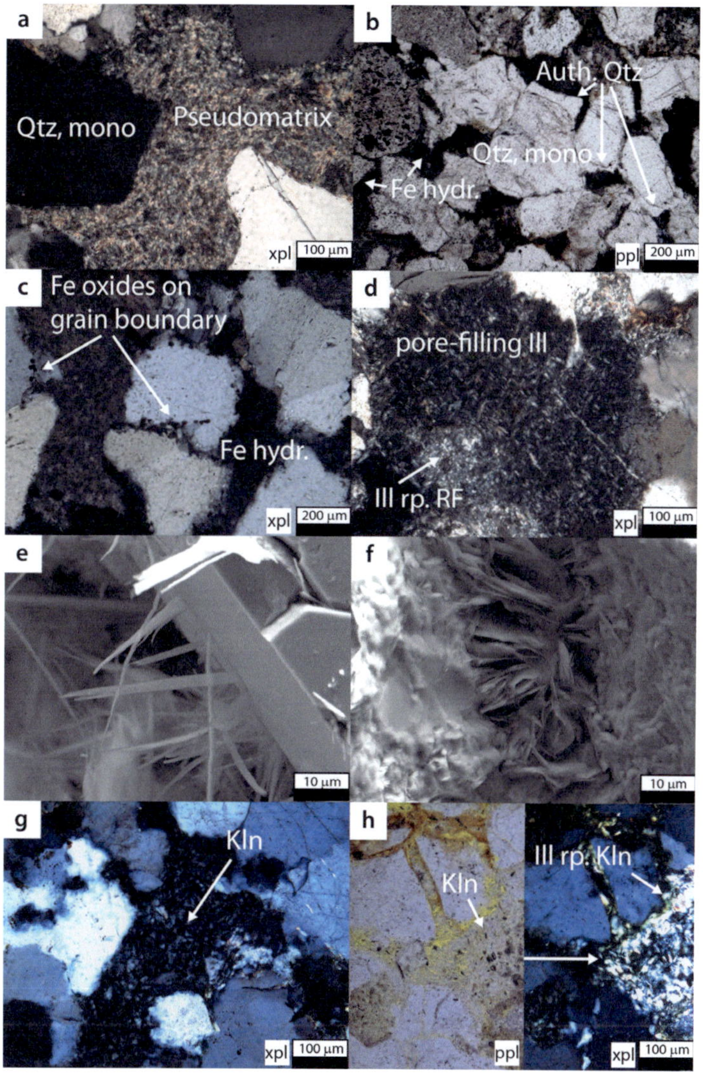

Figure 2.9. Photomicrographs and scanning electron microscopy (SEM) images showing diagenetic characteristics of sandstone samples from Woitzel (Wo) and Hüggel (Hu). a) Ductile grains squeezed between rigid framework grains forming pseudomatrix (sample Wo_028). b) Authigenic quartz forming syntaxial overgrowth on a quartz grain, and iron hydroxides filling pore space (sample Wo_001). c) Red sandstone with iron hydroxide filling the pore space and granular iron oxides at the margins of the quartz grains (sample Hu_07). d) Pore-filling illite replacing a dissolved grain and a rock fragment (sample Wo_030). e) SEM image of illite fibers on framework grains (sample Wo_030). f) SEM image of flakey illite platelets oriented nearly parallel to each other (sample Wo_030). g) Booklet-shaped kaolinite with low interference colors (sample Hu_11). h) Booklet-shaped kaolinite with interparticle microporosity (left) and partial replacement by illite (high interference colors, right) (sample Hu_11). Qtz = quartz, RF = rock fragment, Fe hydr. = Iron hydroxides, Ill = illite, Kln = kaolinite, rp. = replacing, mono = monocrystalline, auth. = authigenic, xpl = cross-polarized light, ppl = plane polarized light.

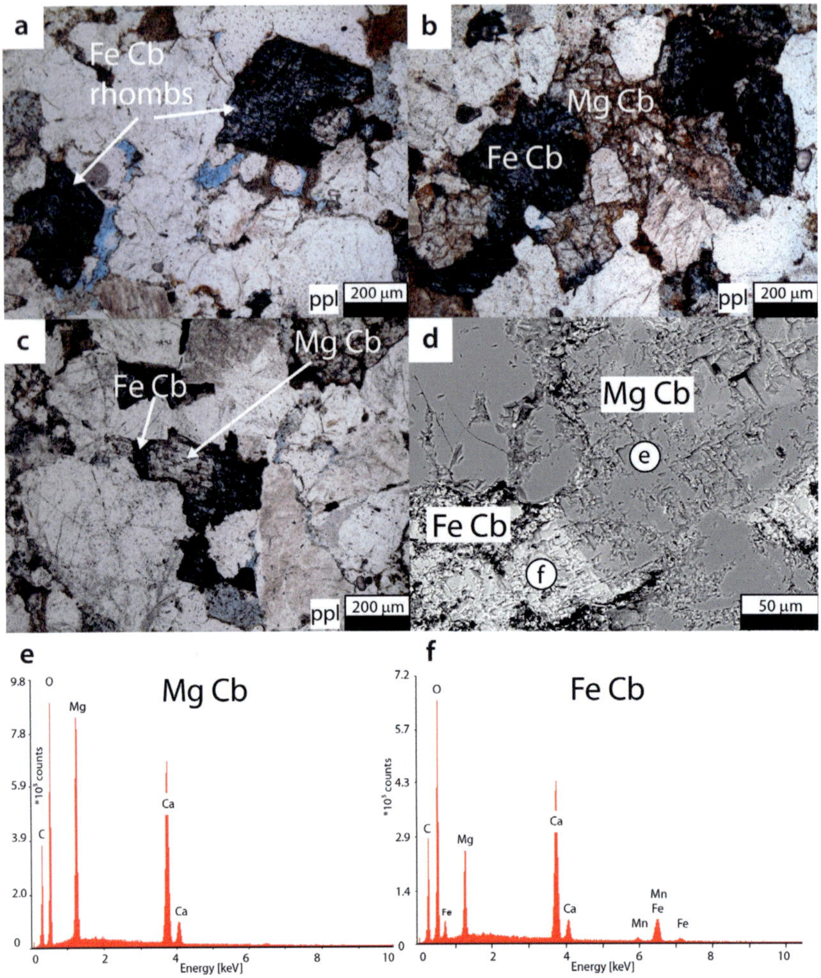

Figure 2.10. Carbonate cement types in the sandstones from Hüggel (blue = ferrous carbonate, unstained = Mg carbonate). a) Ferrous carbonate cement with euhedral rhombohedral crystal shapes (sample Hu_05). b) Ferrous carbonate (blue) and Mg carbonate (colorless) cement types (sample Hu_05). c) Cement zonation with Mg carbonate in the crystal cores and ferrous carbonates in the peripheries. d) The corresponding BSE image shows the two cement phases. e) EDX spectrum of the dark gray carbonate phase showing Ca and Mg contents. f) The lighter gray carbonate phase contains additional Fe. Mg Cb = Mg carbonate, Fe Cb = Fe carbonate, ppl = plane polarized light.

Booklet-shaped kaolinite fills the voids created by dissolved grains (Fig. 2.9g) and is preferentially present near carbonates. Two types of pore-filling kaolinite, one showing partial replacement by illite and one appearing unaffected by illitization, occur in the Hüggel sandstones (Fig. 2.9g and h). Kaolinite booklets unaffected by illitization (Fig. 2.9g) are smaller than those which are partly replaced by illite (Fig. 2.9h).

Table 2.2 (description on next page)

Sample	Detrital components: Quartz	Feldspar	Mica	Accessory Minerals (Ttn, Tur)	Metamorphic RF	Shale RF	Silicic Volcanic RF	Chert	Silicic RF	Coal fragment and Iron oxide	Pseudomatrix	Authigenic components: Quartz cement	Illite pore filling (meshwork)	Illite replaces Feldspar or RF	Illite replaces Kaolinite	Kaolinite	Kaolinite replaces Feldspar	Carbonate cement	Iron oxides/-hydroxides pore filling	Porosity, permeability, and IGV: Secondary porosity (intergranular)	He porosity	Permeability at 10 MPa	Intergranular Volume (IGV)	Grainsize: Average	Maximum	Category	Sorting	Skewness
Wo_001	60.3	1.3	0.0	0.3	8.3	3.0	1.3	0.0	0.3	0.0	1.0	2.0	11.7	7.3	0.0	0.0	0.0	0.0	0.3	2.7	14.3	0.021	16.7	0.3	0.8	medium sand	moderately sorted	coarse skewed
Wo_002	58.3	1.3	0.0	0.0	8.7	3.3	0.7	0.7	0.0	0.0	3.0	0.0	4.7	3.3	0.0	0.0	0.0	0.0	14.0	2.0	13.6	0.006	20.7	0.3	0.7	medium sand	moderately sorted	coarse skewed
Wo_003	51.0	2.0	0.0	0.3	11.3	4.0	0.7	0.7	0.0	0.0	5.7	5.0	8.7	3.0	0.0	0.0	0.0	0.0	4.3	3.3	15.4	0.033	21.3	0.4	0.9	medium sand	moderately sorted	coarse skewed
Wo_004	51.3	2.3	0.7	0.3	14.0	2.3	0.3	1.0	0.0	0.0	4.0	6.7	6.7	2.3	0.0	0.0	0.0	0.0	2.7	5.3	13.6	0.031	21.3	0.4	1.0	medium sand	moderately sorted	coarse skewed
Wo_005	53.7	2.7	0.0	1.3	11.3	3.7	0.7	0.0	0.0	0.0	3.7	3.0	10.3	1.0	0.0	0.0	0.0	0.0	6.3	2.3	15.0	0.025	22.0	0.4	0.7	medium sand	moderately sorted	coarse skewed
Wo_006	60.3	2.0	0.0	1.0	10.3	3.0	0.7	1.0	0.0	0.3	3.7	2.3	7.3	6.0	0.0	0.0	0.0	0.0	0.0	2.0	15.0	0.021	11.7	0.4	0.8	medium sand	moderately sorted	near-symmetrical
Wo_007	52.3	1.7	0.3	0.3	15.0	2.3	0.7	1.0	0.0	0.0	3.0	2.3	6.3	0.7	0.0	0.0	0.0	0.0	12.3	1.7	14.9	0.004	22.7	0.4	0.9	medium sand	moderately sorted	near-symmetrical
Wo_008	54.0	2.7	0.3	0.3	9.0	2.3	1.0	0.0	0.0	0.0	3.3	4.0	8.0	1.0	0.0	0.0	0.0	0.0	6.3	7.7	13.9	0.004	26.0	0.5	0.9	medium sand	moderately sorted	near-symmetrical
Wo_009	60.7	0.3	0.0	0.7	9.0	2.3	1.0	0.0	0.0	0.0	3.3	4.7	4.7	2.7	0.0	0.0	0.0	0.0	9.0	1.7	13.6	0.004	20.0	0.5	1.0	medium sand	moderately sorted	near-symmetrical
Wo_010	61.7	1.3	0.0	1.7	7.3	3.3	0.7	0.3	0.3	0.0	3.3	2.7	7.0	6.3	0.0	0.0	0.0	0.0	1.0	3.0	15.3	0.036	13.7	0.4	0.7	medium sand	moderately sorted	near-symmetrical
Wo_011	55.7	1.3	0.0	1.3	12.7	3.0	0.0	0.3	0.0	0.0	3.7	3.0	7.3	2.3	0.0	0.0	0.0	0.0	5.3	4.0	15.8	0.029	19.7	0.4	1.1	medium sand	moderately sorted	coarse skewed
Wo_012	57.0	2.0	0.0	1.3	8.7	4.3	1.3	0.0	0.0	0.0	5.3	4.3	7.7	3.7	0.0	0.0	0.0	0.0	3.3	1.0	15.2	0.056	16.3	0.4	0.8	medium sand	moderately sorted	coarse skewed
Wo_013	54.7	1.7	0.0	1.0	12.0	3.3	0.0	0.0	0.0	0.0	7.0	4.7	2.3	1.0	0.0	0.0	0.0	0.0	7.7	4.7	13.9	0.042	19.3	0.5	0.9	coarse sand	moderately sorted	coarse skewed
Wo_015	53.7	0.3	0.3	0.7	9.7	3.0	1.7	0.3	0.7	0.0	6.7	4.7	5.7	2.3	0.0	0.0	0.0	0.0	7.7	2.7	14.3	0.003	20.7	0.5	1.7	coarse sand	moderately well sorted	coarse skewed
Wo_018	59.0	1.7	0.0	1.0	10.0	2.0	1.3	0.3	0.0	0.0	2.3	4.3	10.7	2.7	0.0	0.0	0.0	0.0	0.0	4.7	16.3	0.161	19.7	0.6	1.5	coarse sand	moderately well sorted	near-symmetrical
Wo_020	53.3	1.7	0.0	0.0	10.3	1.7	2.7	0.3	0.0	0.0	7.3	2.7	4.7	2.7	0.0	0.0	0.0	0.0	8.7	4.0	15.6	0.037	20.0	0.4	0.8	medium sand	moderately sorted	near-symmetrical
Wo_021	54.3	1.0	0.0	1.7	10.7	4.7	0.3	0.3	0.0	0.3	4.0	2.7	11.3	2.7	0.0	0.0	0.0	0.0	0.3	5.7	19.0	0.091	20.0	0.5	2.4	coarse sand	moderately sorted	fine-skewed
Wo_022	45.7	1.7	0.7	2.0	11.0	3.3	1.7	0.3	0.0	0.0	12.3	1.7	12.3	3.7	0.0	0.0	0.0	0.0	0.3	3.3	18.5	0.014	17.7	0.3	0.5	medium sand	moderately sorted	near-symmetrical
Wo_023	38.0	2.3	2.3	1.3	16.0	6.0	2.3	0.0	0.0	0.0	11.7	1.7	3.0	2.3	0.0	0.0	0.0	0.0	12.0	1.0	17.2	0.029	17.7	0.3	0.6	medium sand	moderately sorted	near-symmetrical
Wo_025	46.3	0.0	0.3	1.0	14.7	3.3	2.0	0.0	0.3	0.3	14.0	4.3	1.3	2.7	0.0	0.0	0.0	0.0	9.0	0.3	15.0	0.010	15.0	0.6	1.1	coarse sand	moderately well sorted	near-symmetrical
Wo_026	59.7	1.7	0.0	0.3	9.0	2.3	0.7	0.0	0.0	0.0	3.7	5.0	7.7	4.3	0.0	0.0	0.0	0.0	2.3	3.3	16.8	0.053	18.3	0.4	0.8	medium sand	moderately sorted	near-symmetrical
Wo_027	59.0	2.3	0.0	0.3	9.0	4.7	0.3	0.7	0.0	0.3	4.3	5.0	6.7	2.7	0.0	0.0	0.0	0.0	0.0	4.7	22.7	0.037	16.3	0.4	0.8	medium sand	moderately sorted	near-symmetrical
Wo_028	47.7	2.3	0.0	1.7	14.0	5.7	0.7	0.0	0.0	0.0	7.0	2.7	9.7	4.0	0.0	0.0	0.0	0.0	0.0	4.7	16.5	0.031	17.0	0.5	1.0	medium sand	moderately sorted	coarse skewed
Wo_029	52.3	2.3	0.0	1.3	11.3	3.3	0.7	0.3	0.3	0.0	8.0	4.7	9.3	3.7	0.0	0.0	0.0	0.0	0.3	2.0	16.7	0.021	16.3	0.4	1.0	medium sand	moderately sorted	coarse skewed
Wo_030	54.0	2.7	0.0	0.7	9.3	4.0	1.7	0.3	0.0	0.0	4.0	7.7	4.0	1.0	0.0	0.0	0.0	0.0	4.7	6.0	16.4	0.015	22.3	0.4	0.8	medium sand	moderately sorted	coarse skewed
Wo_031	52.0	1.7	1.3	0.3	12.7	1.0	0.7	0.3	0.0	0.0	4.7	1.3	4.0	1.7	0.0	0.0	0.0	0.0	14.3	4.0	16.2	0.016	23.7	0.4	1.0	medium sand	moderately sorted	near-symmetrical
Wo_033	48.7	1.0	0.3	0.0	13.3	4.0	1.3	0.0	0.3	0.0	6.3	5.0	2.7	3.0	0.0	0.0	0.0	0.0	13.3	0.7	16.8	0.002	21.7	0.4	0.9	medium sand	moderately sorted	near-symmetrical
Wo_034	51.7	1.3	0.3	0.3	11.7	2.0	0.7	0.0	0.0	0.0	5.0	4.7	7.0	2.3	0.0	0.0	0.0	0.0	9.3	3.7	15.2	0.023	24.7	0.4	1.2	medium sand	moderately sorted	near-symmetrical
Wo_035	47.3	1.3	0.0	0.3	9.0	6.0	1.3	0.0	0.3	2.7	6.0	5.3	5.0	4.3	0.0	0.0	0.0	0.0	4.3	6.7	14.7	0.046	21.3	0.5	1.1	coarse sand	moderately well sorted	coarse skewed
Hu_02	50.0	0.0	0.0	1.0	14.0	1.3	0.0	1.3	0.0	0.0	4.3	5.0	0.0	1.3	0.0	5.3	2.7	0.0	1.3	12.3	-	-	24.0	0.4	0.7	medium sand	moderately sorted	coarse skewed
Hu_03	46.0	0.0	0.3	0.0	14.7	3.7	0.0	0.3	0.3	0.0	6.3	4.3	0.0	0.3	0.0	2.3	0.7	8.3	5.7	6.7	-	-	27.3	0.4	0.8	medium sand	moderately sorted	near-symmetrical
Hu_04	44.3	0.0	0.0	0.0	17.3	3.0	0.0	0.7	0.0	0.3	1.0	5.3	0.0	0.3	0.0	4.7	1.3	0.0	11.3	10.3	-	-	31.7	0.4	0.8	medium sand	moderately sorted	coarse skewed
Hu_05	52.0	0.0	0.0	0.3	9.7	2.7	2.3	0.0	0.3	0.0	3.3	2.0	0.0	0.3	0.0	6.0	0.0	15.7	1.7	3.7	30.9	0.348	29.0	0.4	0.9	medium sand	moderately sorted	near-symmetrical
Hu_06	39.7	0.0	0.0	0.7	9.0	2.7	0.0	0.3	0.0	0.0	35.0	0.7	0.0	0.3	0.0	5.3	0.3	0.0	6.0	0.0	15.5	0.014	12.0	0.2	0.3	fine sand	moderately sorted	coarse skewed
Hu_07	46.7	0.0	0.0	3.7	15.3	4.3	1.3	0.7	0.0	0.7	5.3	0.3	0.0	0.7	0.0	8.7	0.0	0.0	11.3	1.0	17.3	0.033	21.3	0.3	0.5	medium sand	moderately sorted	fine-skewed
Hu_08	61.0	1.0	0.0	0.3	14.0	1.0	0.3	1.0	0.0	0.0	1.3	3.0	3.7	2.3	0.7	4.0	0.3	0.3	0.7	5.0	20.4	2.421	16.7	0.6	1.4	coarse sand	moderately sorted	coarse skewed
Hu_11	55.7	2.3	0.3	0.0	16.7	2.7	0.7	0.0	1.7	0.0	0.7	1.3	0.0	1.3	0.7	8.0	0.7	0.0	2.7	4.7	18.6	0.033	16.7	0.5	2.2	coarse sand	moderately well sorted	fine-skewed
Hu_12	46.7	0.7	1.3	0.3	18.0	1.3	0.7	1.0	0.3	0.0	8.0	1.0	0.3	1.0	0.7	11.3	1.0	0.0	5.3	1.0	15.6	0.472	19.0	0.2	0.6	fine sand	moderately sorted	coarse skewed
Hu_15	57.7	1.0	0.7	0.0	8.7	5.0	0.3	0.3	0.7	0.0	5.3	2.7	1.7	0.7	1.3	6.7	0.0	0.3	4.0	3.0	15.1	0.018	18.3	0.4	1.0	medium sand	moderately sorted	coarse skewed
Hu_17	53.0	0.3	0.0	3.3	7.7	3.3	0.0	0.3	0.3	0.0	6.0	2.7	0.3	0.7	0.0	4.3	0.0	0.0	15.0	2.7	16.8	0.212	25.0	0.3	0.8	medium sand	moderately sorted	near-symmetrical

Table 2.2. (on previous page) Sample information including average point-counted detrital and authigenic compositions (in %), grain size data (in mm), sorting, and petrophysical measurement data (porosity in %, permeability in mD) for plug samples from the Woitzel and Hüggel locations. RF = rock fragment, Ttn = titanite, Tur = tourmaline; silicic RFs include feldspar-rich and sandstone rock fragments.

2.5.3.3 Compaction and porosity

The degree of mechanical compaction can be estimated from the degree of grain deformation and the intergranular volume (IGV), which is the sum of the intergranular porosity, pore-filling cements and depositional matrix (cf. Paxton et al., 2002). The deformation of ductile lithic fragments squeezed between rigid framework grains and the overall average sample IGV of 20.1% (Table 2.2) both indicate that the studied sandstones have undergone intense mechanical compaction. The average IGV values, 19.4% (Woitzel) and 21.9% (Hüggel), respectively, may indicate that the lower Westphalian D sandstones have undergone slightly more intense compaction than the upper Westphalian D sandstones.

Minor pressure dissolution is present in the form of long or concave-convex contacts, and less commonly in the form of sutured contacts associated with IGV values of less than 26% (cf. Paxton et al., 2002).

Due to the high degree of compaction and the presence of authigenic minerals, primary porosity is virtually absent. Secondary porosity, which point counting shows is up to 8% (with an average of 3%) in the Woitzel sandstones and 12% (with an average of 3%) in the Hüggel sandstones, is due to dissolution of detrital and authigenic components (e.g. carbonate cements), as indicated by the rhombohedral pore spaces (Fig. 2.11a). Lithic grains are partly dissolved, and the dissolution of feldspars and aluminosilicates, such as feldspar-rich rock fragments, is suggested by irregularly shaped pore spaces (Fig. 2.11b).

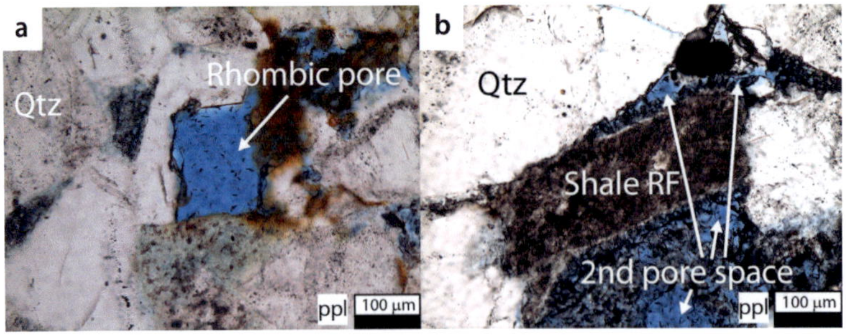

Figure 2.11. a) Secondary pore space resulting from carbonate cement dissolution as indicated by the rhombohedral shape (sample Wo_006). b) Irregular secondary pore space indicates dissolution of additional material such as aluminosilicates or feldspars (sample Wo_010). Qtz = quartz, RF = rock fragment.

The point-counted secondary porosity is lower than the measured He porosity because pore-filling clay phases, such as meshwork illite, contain significant amounts of microporosity which cannot be quantified via optical microscopy. However, the sum of pore-filling illite and/or kaolinite and the point- counted secondary porosity almost equals the measured He porosity (e.g. samples Wo_001 and Hu_12, Table 2.2). If large amounts of iron oxides/-hydroxides are present (as for example in samples Wo_033 and Hu_17; Table 2.2), the oxides/hydroxides are often superimposed on the clay phases, resulting in a difference between the optically-derived and He-measured porosity values due to underestimation of the amount of clay minerals via point counting. However, for quantification issues, the clay micropore distribution would have to be considered and measured in the present samples (cf. Hurst and Nadeau, 1994). Overall, if the difference between the point-counted and He-measured porosity only depends on microporosity, microporosity appears to account for almost 80% of the total porosity present in both Woitzel and Hüggel samples.

2.5.4 Petrophysics

The porosity and permeability values of the 33 sandstone samples from the vertical and horizontal profiles of the lower Westphalian D sandstones in the Woitzel quarry, and of the 15 samples from the upper Westphalian D sandstones in the Hüggel area, were measured. In both sets of sandstone samples, there is a correlation between porosity and permeability (Fig. 2.12).

For the Woitzel samples, the mean permeability is 0.03 mD (± 0.005 mD, statistical error) for a confining pressure of 10 MPa and 0.017 mD (± 0.003 mD) for a confining pressure of 30 MPa. Therefore, an increase in the confining pressure of 20 MPa reduces the permeability by a factor of 0.57, resulting in an average pressure sensitivity coefficient of 0.03 MPa^{-1}. These samples have a mean porosity of 15% (± 0.3%) with negligible variation at different confining pressures (Fig. 2.12).

For the Hüggel sandstones, the permeability values have a mean of 0.544 mD (±0.16 mD) at a confining pressure of 10 MPa (Fig. 2.12). The permeability is reduced by a factor of 0.67 with an increase in the confining pressures to 30 MPa. This compaction-induced permeability reduction results in a pressure sensitivity coefficient with an average of 0.03 MPa^{-1}. The most permeable samples, with permeabilities on the order of 1 mD, appear to be less affected by increases in the confining pressure. The Hüggel sandstones have a mean porosity of 19% (±1%).

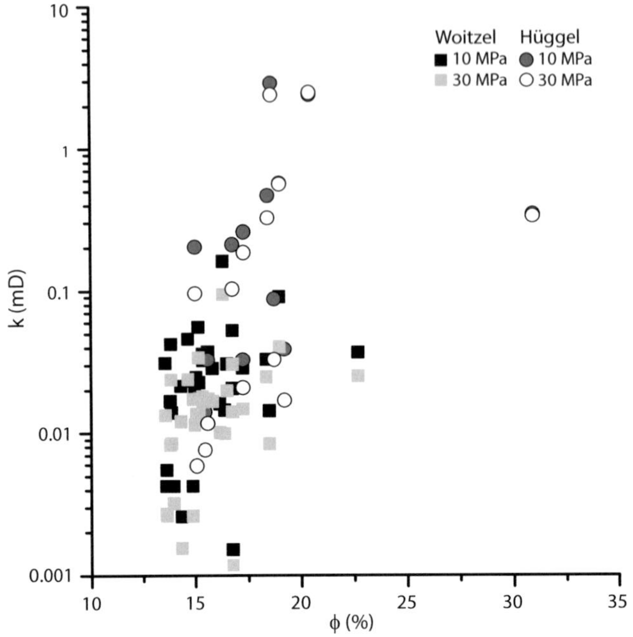

Figure 2.12. Permeability-porosity cross-plot of 33 sandstone samples from Woitzel at 10 MPa (black squares) and 30 MPa (gray squares) confining pressure, and of 15 sandstone samples from Hüggel at 10 MPa (gray circles) and 30 MPa (white circles). The Woitzel samples have porosity values of approximately 15% and very low permeability (0.001 – 0.1 mD). In the Hüggel samples, permeabilities are up to 1 mD. Samples from both outcrops tend to show lower permeability values at higher confining pressures with the exception of three Hüggel samples (Hu_05, 08 and 11). For these samples, permeability values remain almost the same despite the higher confining pressure.

Grain size variations may influence reservoir characteristics. The combined set of Woitzel and Hüggel sandstone samples exhibits no general correlation between permeability and grain size, as indicated by the low correlation coefficient ($R^2 = 0.12$) (Fig. 2.13). Regarding each studied area separately, the Hüggel sandstones show a distinct correlation between permeability and grain size ($R^2 = 0.66$), whereas the Woitzel samples show no such correlation ($R^2 = 0.06$). However, if the data set is restricted to Woitzel samples with permeabilities greater than 0.01 mD, a correlation is observed ($R^2=0.52$) (Fig. 2.13). By contrast, no correlation is observed for either set of samples between porosity and grain size (Fig. 2.13).

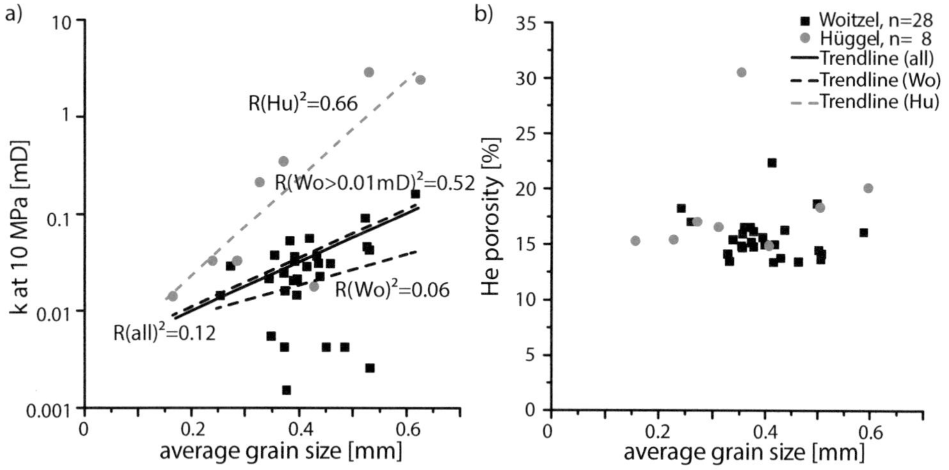

Figure 2.13. a) Cross-plot between permeability and average grain size for samples of the Hüggel (Hu) and Woitzel (Wo) sandstones. Samples of upper Westphalian D sandstones from Hüggel ($R^2 = 0.66$) show some correlation with grain size, as do lower Westphalian D sandstones from Woitzel with k>0.01 mD ($R^2 = 0.52$). b) Cross-plot showing that there is no correlation between porosity and average grain size for sandstone samples from Hüggel and Woitzel.

2.5.4.1 Permeability and diagenetic minerals

The permeability was plotted against the proportions of pore-filling quartz, illite, iron oxides/-hydroxides, and pseudomatrix for all samples based on the associated thin sections (Fig. 2.14).

The results show that there is no correlation between permeability and authigenic quartz and illite for the Woitzel and Hüggel sandstones, as evidenced by the very low correlation coefficients ($R^2<0.05$). If the data set is restricted to samples with permeabilities greater than 0.01 mD, the correlation coefficients are slightly higher ($R^2 = 0.1$). For samples with low cement proportions (below 5%), no control on permeability is observed, as the values vary over several orders of magnitude (Fig. 2.14). In addition, the correlation with iron oxides/-hydroxides shows no distinct trend ($R^2= 0.16$) (Fig. 2.14). The amount of pseudomatrix is negatively correlated with permeability in the Hüggel sandstones ($R^2=0.75$), but not correlated with permeability in the Woitzel sandstones ($R^2 = 0.001$) (Fig. 2.14).

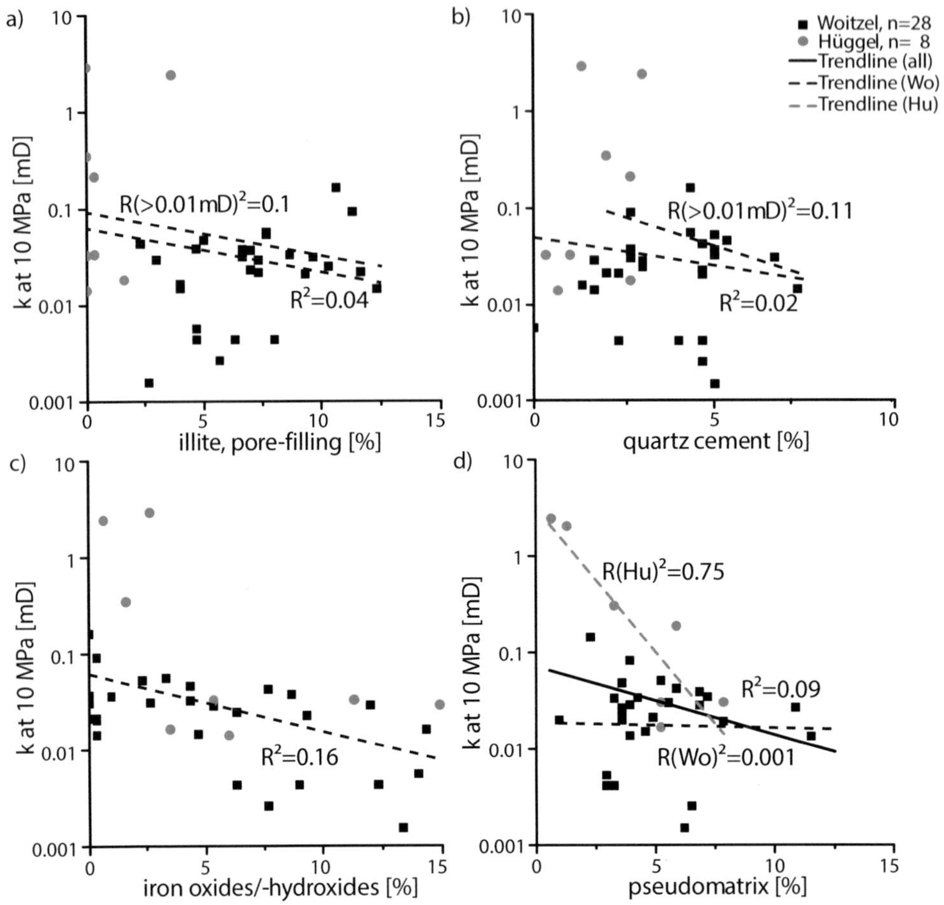

Figure 2.14. Cross-plots of permeability (mD) versus pore-filling a) illite, b) quartz, c) iron oxides/-hydroxides, and d) pseudomatrix for the sandstone samples from Hüggel and Woitzel. The plots show that there is little correlation with permeability, except for pseudomatrix in the Hüggel sandstones (R^2=0.75).

2.5.4.2 Vertical and lateral trends

The vertical section in Woitzel Quarry 2 (Profile a, Figs. 2.2a and 2.3) comprises 11 samples spanning 37 m of stratigraphy (Fig. 15a, b). These sandstones have a mean porosity of 16% (± 0.6%) (Fig. 2.15c) and permeability values which range from 0.001 mD to 0.1 mD (Fig. 2.15c). The low-permeability sample Wo_015 (k<0.01 mD) exhibits no relationship between permeability and grain size. The presence of significant amounts of quartz cement, illite, and iron oxides/-hydroxides in this sample may contribute to its low permeability (Fig. 2.15c).

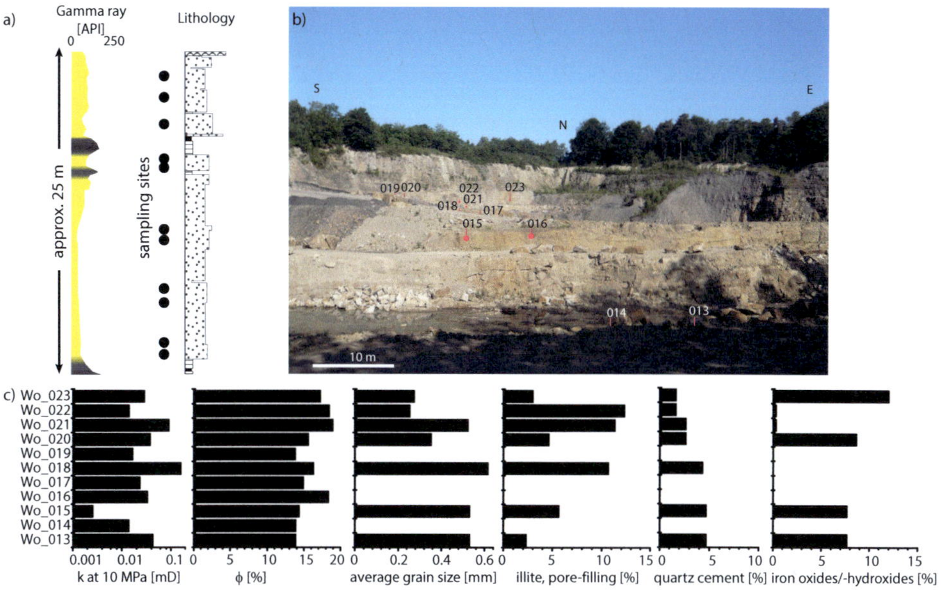

Figure 2.15. a) Vertical lithological profile of the quarry wall at Woitzel Quarry 2 (profile a, Fig. 2a) showing sample positions, gamma-ray profile and lithological column; b) photograph of Woitzel "Quarry 2". c) Bar charts showing vertical variations in permeability, porosity, grain size, illite (%), quartz content and Fe-oxide/hydroxide content in the analyzed samples (Wo_013 to Wo_023). Trends in permeability and porosity show little relationship with variations in lithology. Grain size variations in general correspond to permeability variations, with the exception of sample Wo_015 which has a large average grain size but a low permeability; however, large amounts of cement and authigenic illite are present in this sample.

A lateral profile of 21 samples spanning a distance of 150 m was taken from a single medium-grained sandstone unit below the Alexander coal seam in the Woitzel quarry (Fig. 2.16a). These samples have a mean porosity of 15.5% (± 0.5%), with a maximum of 22% in one sample (Wo_027). Despite this slightly higher porosity, no petrographic differences occur between the samples (see Table 2.2). Permeabilities range from 0.001 mD to 0.1 mD (Fig. 2.16b). The correlation between permeability and grain size is not strong. In addition, the results show little correlation between permeability and the amount of diagenetic quartz or illite. However, the amount of late diagenetic iron oxides/-hydroxides is negatively correlated with the permeability (e.g. Wo_033) (Fig. 2.16b).

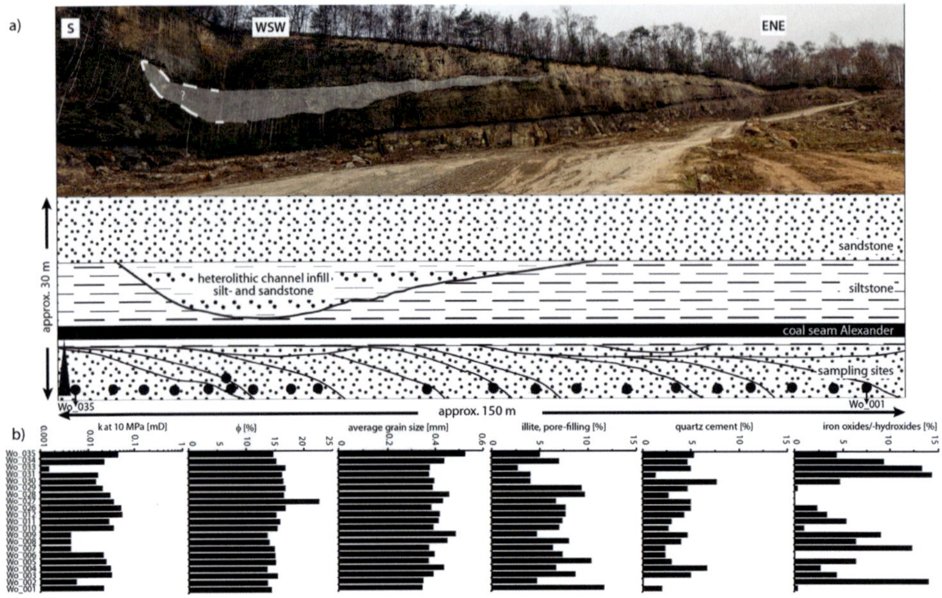

Figure 2.16. a) Field photograph of the northern wall of the Main Quarry at Woitzel (location in Fig. 2a) (above) with schematic interpretation showing the sedimentary structure (below). Permeability and porosity were measured on 21 samples covering the entire profile length of more than 150 m. The sampled unit comprises a series of laterally-accreting point bar sandstones which internally show fining-upward grain-size variations. b) The bar charts show that while sample porosities remain almost constant at 15% with one outlier (>20%, Wo_027), permeabilities range over almost two orders of magnitude from 0.001 to 0.1 mD. Permeability variations cannot be correlated with grain size variations or amounts of authigenic cement; however, higher contents of iron oxides/-hydroxides correspond to low permeability values.

2.6 Discussion

The sandstones in the Woitzel quarry represent several fining-upwards sequences in a multi-storey, braided to meandering fluvial system with intercalated coal seams (see Figs. 2.3 and 2.4d). The compound successions of individual, meter-thick bar deposits with low-angle cross-stratification and minor conglomeratic intervals are indicative of an overall relatively low-energy depositional environment. Bar deposits with downstream accretion surfaces occur in the lower parts of the Woitzel quarry. In the upper part of the quarry, current ripples at an angle to the larger-scale accretion surfaces indicate that the sandstones represent point bar deposits (see Fig. 2.4e and f). In the Main Quarry, a several-meter thick siltstone succession (see Fig. 2.3) with thin sandstone intercalations overlies the Alexander coal seam and represents deposition in a floodplain environment with crevasse splay deposits resulting from local flooding events. The large channel incision in the Main Quarry (see Fig. 2.5) represents an abandoned oxbow deposit in this silty floodplain environment. Outcrop gamma ray logging points to fining-upwards fluvial cycles (see Fig. 2.3), with low gamma ray values corresponding to sandy units and higher values corresponding to units with high clay contents (cf. Rider and Kennedy, 2011).

During Westphalian D time, the climate changed from humid to more arid, resulting in a decrease in the occurrence of coal seams throughout the Lower Saxony Basin (Besly et al., 1993; David, 1990; Jones and Glover, 2005). The climate change is also recorded in the transition from gray to red sandstones, which occurs in an interval 134 m above the Johannisstein coal seam observed in a coal mining shaft in the Hüggel area (Fig. 2.17) and which is overlain by 355 m of red siliciclastics (Klassen, 1984). The upper Westphalian D sandstones from Hüggel are therefore interpreted to have been deposited under arid climatic conditions, while the lower Westphalian D sandstones from Woitzel are interpreted to have been deposited under more humid conditions. This occurrence of red bed deposits in the uppermost Westphalian sequences is known throughout NW Europe, and especially in the southern North Sea, where they contain large hydrocarbon reserves (Besly et al., 1993; Leeder and Hardman, 1990; Moscariello, 2003). Therefore, a detailed study of the present two outcrops may contribute to the better understanding of the reservoir characteristics of this red bed succession.

Both gray and red sandstones are present in the Hüggel outcrop (see Fig. 2.6). Gray sandstones occur as a 5 m thick interval directly beneath the Zechstein unconformity (cf. Klassen, 1984). The unconformity was identified from the outcrop by the presence of the overlying copper shales and can be confirmed by logging due to their high gamma-values (see Fig. 2.3) (cf. Rider and Kennedy, 2011).

In the gray sandstone intervals, processes similar to the Weissliegend or bleached Rotliegend intervals most likely affected the lithology and likely included the leaching of hematite. Some authors (e.g., Gaupp and Okkerman, 2011) assume that these processes were related to the degassing of the Westphalian coal seams, which released hydrocarbons and carbon dioxide-bearing formation waters, resulting in a decrease in pH. The acidic fluids ascended until they were stopped by the sealing copper shale resulting in leaching of the sandstones below this barrier. In addition, thermal maturation of the copper shale causes an acidic and reducing environment with bleaching capacity. Other authors (e.g., Glennie and Buller, 1983; Peryt et al., 2010) proposed that the absence of red color due to the anoxic seafloor environment during the Zechstein transgression. The presence of reworking features, such as soft-sediment deformation or bioturbation on the seafloor, could support the latter model but were not observed in the Hüggel outcrops. The locally exposed red sandstones show cross bedding, pebbly intervals, and a lack of rootlet or soil features, suggesting the presence of meandering channels on a well-drained floodplain under arid conditions (Besly and Fielding, 1989; Bridge, 2006).

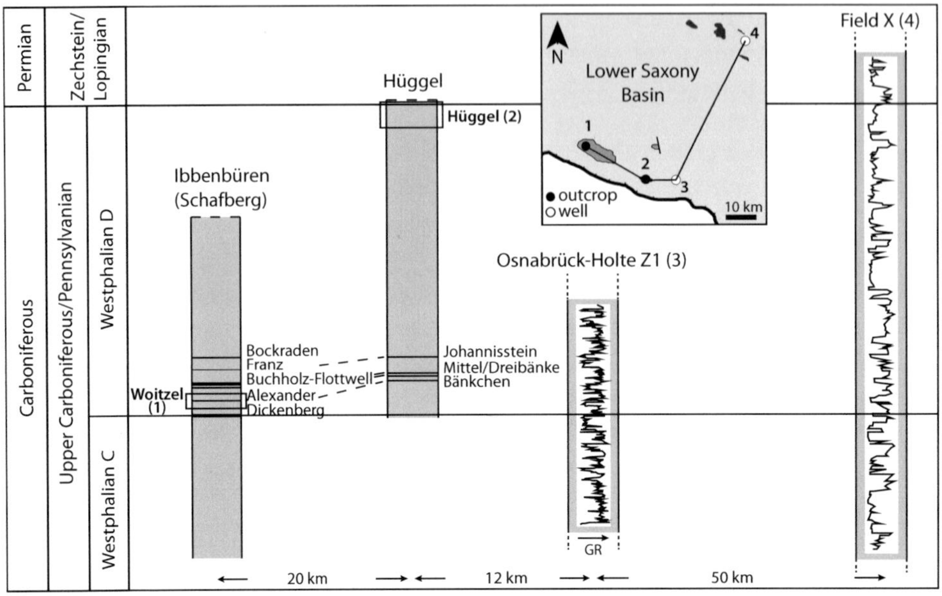

Figure 2.17. Correlation of the Westphalian C/D boundary between outcrops in the south of the Lower Saxony Basin (columns 1, 2) and the subsurface stratigraphy at a nearby borehole (column 3) and at field X to the north (column 4) (modified after Jones and Glover, 2005; Klassen, 1984), together with the correlation of the coal seams in the Schafberg-Hüggel area (modified after Klassen, 1984). The Dickenberg coal seam marks the Westphalian C/D boundary. Above the boundary, the basal Westphalian D sandstones are gray in color and intercalated with numerous coal seams of variable thickness. Coals are not present in the upper Westphalian D and sandstones show red coloration (hatched area) indicating a change in climate towards arid climatic conditions. The gray/red boundary at the Schafberg location (see Fig. 1b) is located 200 m above the Buchholz-Flottwell coal seam and is overlain by some 350 m of red beds (Drozdzewski, 1985), which are not exposed in the Woitzel quarry. At the Hüggel locations, the transition to red colored deposits occurs at a depth of 410.6 m, which is 134 m above the Johannisstein coal seam (Klassen, 1984).

2.6.1 Paragenetic sequence

The terms eo-, meso-, and telodiagenesis have been suggested by Choquette and Pray (1970) and Schmidt and McDonald (1979) to address the effects of meteoric waters on diagenesis. Alternatively, Clark (1986), for example, used the terms surface, shallow burial, deeper burial and uplift diagenetic phases to describe the burial stages. Here, the paragenetic sequence of the studied sandstones is divided into early, burial, and uplift diagenetic phases. Early diagenesis occurs in an environment where diagenetic reactions are in general controlled by surface conditions; burial diagenesis is defined as occurring at temperatures greater than 70°C as numerous processes, such as chemical compaction, clay-mineral transformation reactions and burial quartz cementation begin to occur at this temperature (Morad et al., 2000). All processes active during the basin inversion are referred to collectively as uplift diagenesis.

In the following section, these stages of diagenesis are related to the burial history of the Upper Carboniferous rocks in the Lower Saxony Basin as reconstructed by Bruns et al. (2013) (Fig. 2.18). These authors recognized an initial burial stage during the Late Carboniferous followed by uplift during the Permian before the main phase of burial occurred until the Late Cretaceous. Subsequent Late Cretaceous and later inversion caused rapid uplift and the sandstones were finally exposed at the surface during the Quaternary (Senglaub et al., 2005a).

The main similarities and differences in the authigenic composition of the Westphalian D sandstones in the study areas can be summarized as follows:

- Iron oxides and hydroxides are common in the Woitzel and Hüggel sandstones, whereas oxides with a granular appearance arranged on the rims of framework grains are restricted to samples from Hüggel (Fig. 2.9b and c).
- Similar amounts of quartz cements are present in all the sandstone samples from both locations (Table 2.2, Fig. 2.9b).
- Two phases of kaolinite are present in the Hüggel sandstones, one of which appears to be partially illitized (Fig. 2.9g and h). No kaolinite is present in the Woitzel sandstones.
- Illite is common in the Woitzel sandstone samples, but occurs only in small quantities in some Hüggel samples (Fig. 2.9d-f, h).
- Carbonate cement is only preserved in some samples of Hüggel sandstones and is represented by Mg-rich and Fe-rich phases (Fig. 2.10).

2.6.1.1 Early diagenesis I

Early diagenetic processes include the infiltration of clay minerals, iron oxides and – hydroxides, which precipitate from percolating mud-rich surface waters (Worden and Morad, 2003). In the sandstone samples from the Woitzel quarry, no unambiguous evidence for early clay infiltration is observed. In the Hüggel sandstones, granular iron oxides located on quartz grain boundaries and partly overgrown by authigenic quartz indicate early formation in an arid environment or formation during later uplift and potential exposure under arid Permian conditions (see Fig. 2.9c). This matches with observations from the Westphalian C and D red beds, which Besly et al. (1993) suggested were reddened syn-depositionally.

Other diagenetic processes during early burial include the dissolution of feldspars and the precipitation of kaolinite (Ketzer et al., 2003; Lanson et al., 2002; Morad et al., 2010), due to the presence of percolating undersaturated meteoric waters (Lanson et al., 2002; Morad et al., 2000; Worden and Morad, 2003). Therefore, kaolinite most likely was formed under these conditions in the Woitzel sandstones. However, kaolinite is no longer present in the studied samples, indicating the replacement of the early-formed kaolinite during later burial. In contrast, in the upper Westphalian D sandstones at Hüggel, arid climatic conditions prevailed during early burial, and early kaolinite formation processes were potentially slower, resulting in a smaller volume of early-formed kaolinite.

Incipient quartz overgrowth development during early burial was observed in the Westphalian sandstones by Sedat et al. (1992) but could not be confirmed in the study area.

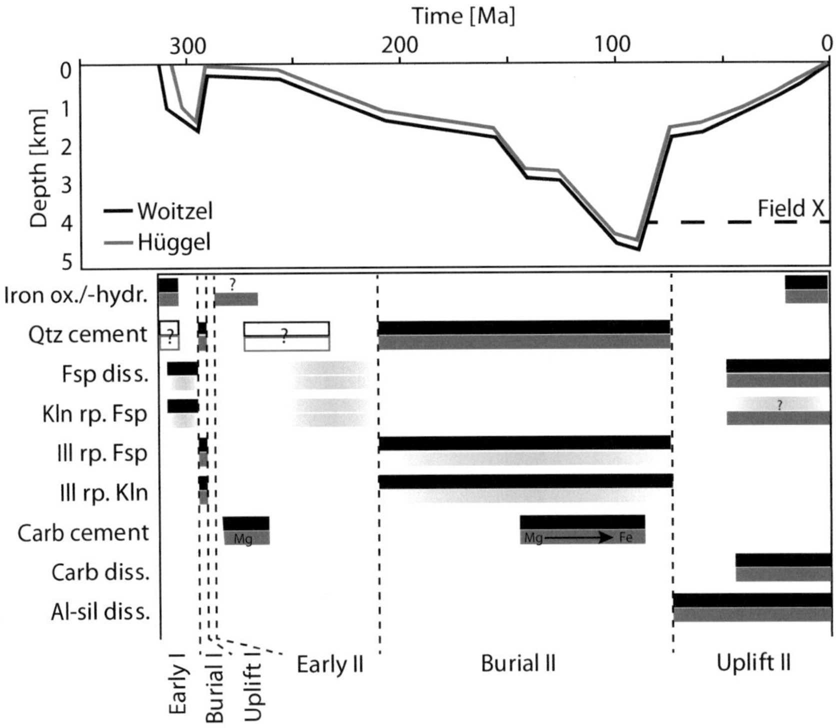

Figure 2.18. Paragenetic sequence for the Westphalian D sandstones correlated with the burial history at Woitzel and Hüggel (modified after Bruns et al., 2013) and the current depth of a reservoir at a tight gas field in the Lower Saxony Basin for comparison (Field X; Wimmers and Koehrer, 2014) (Locations in Fig. 2.1b). The paragenetic histories of the Woitzel (black) and Hüggel (gray) sandstones are compared to show their different mineral assemblages. Possible mineral phases, which may be expected to occur given the prevailing diagenetic conditions but which have not actually been observed including early quartz and late kaolinite, are labelled with question marks.

2.6.1.2 Burial diagenesis I

Together with the early burial reactions and the pore water chemistry, increases in the temperature, the pressure, and the duration of these parameters drive deeper burial processes (e.g., Füchtbauer, 1988; Morad et al., 2000; Worden and Morad, 2003). In the Late Carboniferous, the studied sandstones are modelled to have been buried to depths of about 2 km, and the associated temperatures (cf. Bruns et al., 2013) were sufficiently high for illitization and quartz cementation to occur. However, no clear evidence for this first phase of burial diagenesis can be distinguished petrographically.

2.6.1.3 Uplift diagenesis I

During the Permian, an initial uplift stage from a depth of approximately 2 km to a depth of 100 m occurred (Bruns et al., 2013) under an arid climate (Fig. 2.18). The prevailing arid conditions probably ruled out the major influence of dissolution processes during the first stage of uplift, but this cannot be confirmed based on the petrographic data from the study area.

2.6.1.4 Early diagenesis II

The semi-arid conditions in the Permian, together with evaporitic pore waters, induced the formation of carbonate cements (Morad et al., 2000; Sedat et al., 1992; Ziegler, 2006). Direct contact with the Zechstein deposits resulted in the presence of saline fluids. Staining and EDX analysis showed the presence of a carbonate cement phase within the Hüggel sandstones, and the partly euhedral shapes of these mineral grains, which are hosted within syntaxial quartz cement overgrowths, confirm that they formed during an early stage (see Fig. 2.10a and b).

Arid conditions also favor the formation of iron oxides and hydroxides, as observed in the Hüggel samples (cf. Worden and Morad, 2003) (see Fig. 2.9c). However, these minerals may alternatively have formed shortly after deposition (see chapter 2.6.1.1).

The processes of kaolinite formation and initial quartz cementation (as described in 'Early diagenesis I') could have occurred, but only small proportions of kaolinite may have formed due to the arid conditions (Bjørlykke, 1998; Hancock and Taylor, 1978; Ketzer et al., 2003).

2.6.1.5 Burial diagenesis II

The major burial related processes occurred during the Jurassic and Cretaceous. Studies on the grade of coalification by Klassen (1984) and Teichmüller (1986) suggest that the succession at Woitzel experienced higher temperatures (coals of low anthracite rank) than that at Hüggel (coals of only bituminous coal rank). The two outcrops are separated by a stratigraphic interval of approximately 600 m (Fig. 2.3) (Drozdzewski, 1985); thus, their different grades of coalification are consistent with the expected grade of coalification at the respective relative depths in NW Germany as proposed by Teichmüller (1986).

Temperature is one controlling factor when considering illite formation. According to many authors (e.g., Bjørlykke and Egeberg, 1993; Worden and Burley, 2003), illite formation starts at temperatures above 70°C and is most pervasive when temperatures reach 130°C. Thus, the temperatures of the rocks in both study areas would have been sufficiently high for intense illitization to occur. However, pervasive illitization is only recorded in samples from Woitzel. In contrast, in samples from Hüggel, few kaolinite booklets show illite fibers indicating either overgrowth or replacement by illite (see Fig. 2.9). Illite can form e.g. by replacing feldspars (see Fig. 2.8a) and kaolinite (e.g., Bjørlykke and Aagaard, 1992; Bjørlykke and Egeberg, 1993; Chuhan et al., 2001; Lanson et al., 2002), which explains the absence of kaolinite in the Woitzel sandstones. In samples from Hüggel, a smaller amount of potentially replaceable kaolinite was present due to the arid conditions during and after deposition. However, other authors (Ehrenberg, 1991; Gaupp et al., 1993; Thyne, 2001) suggest that the illite/kaolinite distribution is primarily controlled by the availability of potassium (high K+/H+ ratio) which is mainly derived from feldspar dissolution or influx of K+-rich waters, rather than as a result of temperature. In the vicinity of organic-rich shale layers, acidic conditions cause a reduction of the K+/H+ ratio, favoring kaolinite preservation (cf. Van Keer et al., 1998). Thus in addition to the reduced availability of kaolinite, the acidic environment in the upper Westphalian D Hüggel sandstone unit below the copper shale may have inhibited illite formation compared to the lower Westphalian D Woitzel sandstones.

The illitization process produces an excess of quartz (e.g., Bjørlykke and Aagaard, 1992; Bjørlykke and Egeberg, 1993; Chuhan et al., 2001; Lanson et al., 2002) leading to the formation of authigenic quartz overgrowths during burial, which are present in samples from both studied areas (see Table 2.2, Fig. 2.9b).

A late phase of carbonate cementation occurred in the second burial stage, partly replacing pre-existing dolomite that formed during the first uplift phase. This later phase is recorded in the Hüggel sandstones (see Fig. 2.10) and is enriched in Mg, Fe, and Mn, potentially sourced from smectite in adjacent mudstones or altered rock fragments. If large amounts of Fe are supplied, ankerite is likely to form under the same conditions (Boles and Franks, 1979; Gaupp et al., 1993; Ziegler, 2006) and is common in Upper Carboniferous sandstones (e.g., Besly et al., 1993; Moscariello, 2003; Sedat et al., 1992). However, ankerite is not present in the studied rocks.

2.6.1.6 Uplift diagenesis II

The second major phase of uplift from a depth of approximately 5 km to the surface began at the end of the Cretaceous (Bruns et al., 2013) (Fig. 2.18). Uplift-related diagenetic processes are pervasive at depths shallower than 1.5 km, and the study region has been exposed to humid climatic conditions since the Paleogene. Processes associated with humid climates include the dissolution of detrital aluminosilicates and related kaolinite precipitation. This late uplift-related kaolinite is generally discussed to be associated with dissolved carbonates related to earlier diagenetic stages (Morad et al., 2000). Due to the relatively short time spent at shallow depth kaolinite booklets appear to be smaller and more disordered than burial-related kaolinite booklets (cf. Beaufort et al., 1998; Lanson et al., 2002). A second kaolinite phase with small booklets (see Fig. 2.9g) thus likely occurred in the Hüggel sandstones, explaining why the majority of the observed kaolinite appears to be unaffected by illitization (see Fig. 2.9g and h). No clear evidence of a second kaolinite phase is recorded in samples from Woitzel. The intense illitization may have made the rocks too tight for sufficient fluid flow to occur.

The general enhancement of porosity by the dissolution of aluminosilicates or carbonate cements is due to infiltrating meteoric waters, and is commonly observed in humid climates (e.g., Ketzer et al., 2003; Lanson et al., 2002). Undefined secondary pore space (due to aluminosilicate dissolution) and rhombohedral pore shapes suggesting the presence of former carbonate cements (see Fig. 2.11) are present in samples from both study areas. The dissolution process, which continues to be active because the sandstones are close to the surface, has played an important role in the increase in total porosity.

Bacterial degradation of organic material generates reducing conditions, which may have led to the mobilization of both iron and manganese phases (Sedat et al., 1992) during the late second uplift stage. Both oxide phases are known to be localized along fractures and are often associated with Liesegang weathering (Chan et al., 2000). The presence of these phases as iron hydroxides and manganese oxides were commonly observed in the sandstone samples from the Woitzel and Hüggel sites (see Figs. 2.4b, 2.6c, 2.9b).

2.6.2 Factors controlling reservoir quality

The reservoir quality of the lower Westphalian D Hüggel sandstones (average permeability: 0.54 mD; average porosity: 19%) is higher by one order of magnitude in terms of permeability than that of the Woitzel sandstones (average permeability: 0.03 mD; average porosity: 15%). These results can be compared to reservoir data from the southern North Sea (UK), where mean porosities of 11-13% (channel sandstones) and a wide range of permeabilities (0.001 to >100 mD) are observed depending on the precise facies (Moscariello, 2003). Besly et al. (1993) noted that the reservoir quality is higher in the Westphalian C red beds than in the gray, coal-bearing Westphalian B sandstones of the UK offshore sector 44. Reservoir properties are variable in the study areas, as shown by the lateral and vertical sections in the Woitzel quarries (see Figs. 2.15 and 2.16). Considering that permeability values vary by two orders of magnitude (0.001 to 0.1 mD) vertically over 25 m (Fig. 2.15) and laterally over 150 m (Fig. 2.16) in a single fluvial sandstone bed, the extrapolation and upscaling of petrophysical well data remains challenging. Similar variations in reservoir quality within a single well and over short inter-well distances of 400 m were observed in the tight gas sandstones with permeability below 0.1 mD in the Lower Saxony Basin (cf. Well A and B in Wimmers and Koehrer, 2014).

Structural effects can result in increased fracture permeabilities. As neither study area has experienced intense fracturing or faulting, these effects are neglected in this study. However, the effect of uplift related microfracturing can be reduced by using increased confining pressures (cf. Amann-Hildenbrand et al., 2016; David et al., 1994). Thus, to ensure comparability of the two data sets from Woitzel and Hüggel, permeabilities measured under different confining pressures were recorded to exclude the influence of potential fracture permeability (see Fig. 2.12).

2.6.2.1 Effects of grain size and depositional environment on ϕ and k

The depositional system can control grain size variations in fluvial sandstones (cf. Bjørlykke, 1998; Olivarius et al., 2015). The lower Westphalian D (Woitzel) sandstones and upper Westphalian D (Hüggel) sandstones both have average grain sizes of 0.4 mm. Therefore, grain size variations cannot control all the recorded variations in reservoir characteristics (see Fig. 2.13).

However, for the kaolinite-bearing upper Westphalian D sandstones, grain size correlates with permeability (see Fig. 2.13a, R^2=0.66). For the illite-bearing lower Westphalian D sandstones, grain size correlates with permeability for rocks with permeabilities greater than 0.01 mD (see Fig. 2.13a, R^2=0.52). Observations of vertical variations on the order of tens of meters within these cyclic fluvial deposits confirm this controlling factor by exhibiting a positive correlation between permeability and grain size (see Fig. 2.15), with a cut-off of 0.01 mD for the grain size effect (see Fig. 2.15). In terms of lateral extent, the sedimentary structures observed are interpreted as separate point bars with lateral accretion bodies and ripples with foresets showing internal grading from coarser- to finer sand grain-sizes (see Figs. 2.4e,f and 2.16), and consequently exhibit small-scale depositional heterogeneity. These sets generally contain slightly more permeable coarse-grained layers (see Fig. 2.4e) which can be linked to the base of the individual foresets at a meter scale. This inherent depositional heterogeneity within a individual fluvial sandstone units should be taken into account in reservoir modeling studies of analogous settings, because the use of a lateral grid with cell dimensions that are too coarse (e.g. 50 x 50 m) may not correctly average the permeability (Fig. 2.19).

2.6.2.2 Effects of depositional environment and diagenesis on ϕ and k

The sediment provenance and climatic conditions at the time of deposition influence the original composition of a sandstone, which is then affected by near-surface diagenetic processes (cf. Khalifa and Morad, 2015; Kim et al., 2007; Morad et al., 2010). These factors and processes can influence the reservoir quality. In the sandstones investigated in this study, the deformation of ductile grains due to mechanical compaction has reduced permeability and primary porosity (cf. Leeder and Hardman, 1990; Molenaar et al., 2015). Pseudomatrix occurs in all the studied samples, but does not appear to exert a general control on the permeability based on all the data. However, the permeability of the upper Westphalian D sandstones is negatively correlated with the pseudomatrix content (see Fig. 2.14d, R^2 = 0.75). Some samples with unusually high pseudomatrix contents have especially low permeabilities (see Table 2.2, e.g. Hu_06: 35% pseudomatrix, 0.01 mD). Therefore, the presence of pseudomatrix locally contributes to variations in reservoir characteristics but does not control the general permeability trend from the lower to the upper Westphalian D sandstones.

The original feldspar content affects the reservoir storage potential, as dissolution and associated authigenic mineral formation can create isolated microporosity (cf. Bertier et al., 2008; Gaupp and Okkerman, 2011; Ketzer et al., 2003; Molenaar et al., 2015). At 4.7%, the original feldspar content of the lower Westphalian D sandstones in the Woitzel quarries is higher than that of the upper Westphalian D sandstones in the Hüggel area, with an original feldspar content of 1.5% (see Table 2.2).

Early kaolinite formation requires a K-feldspar source and the influx of meteoric waters (cf. Lanson et al., 2002; Morad et al., 2000). Due to the higher original feldspar content and humid climate conditions of the lower Westphalian D sandstones, more early kaolinite formed in the Woitzel sandstones than in the upper Westphalian D Hüggel sandstones. The effects of climate and original composition therefore provide a first-order explanation for differences in clay cementation between the two locations.

2.6.2.3 Effects of diagenesis on ϕ and k

Observations in this study indicate that illite is more abundant in the Woitzel sandstone samples than in the Hüggel sandstones. Illite formation requires a source for potassium and aluminum which is likely provided by feldspar and kaolinite (Bjørlykke and Aagaard, 1992; Bjørlykke and Egeberg, 1993; Chuhan et al., 2001; Lanson et al., 2002). Thus intense illitization occurred in the Woitzel sandstones but only minor illite replacement occurred in the Hüggel sandstones, thereby explaining the better reservoir characteristics of the latter units. In addition, the presence of the organic-rich copper shale in the Hüggel location created an acidic subsurface environment with low K^+/H^+ ratios, and prevented the replacement of kaolinite by illite (cf. Gaupp et al., 1993; Thyne, 2001; Van Keer et al., 1998).

Similar intensities of quartz cementation were observed in samples from both study areas (see Fig. 2.14). Therefore, this process does not explain the differences in reservoir characteristics.

The humid climatic conditions during the second uplift stage led to the infiltration of meteoric water, which enhanced the permeability of the sandstones by dissolving unstable carbonate cements and aluminosilicates, such as feldspars, and resulted in precipitation of the main phase of kaolinite in the Hüggel sandstones (cf. Ketzer et al., 2003; Lanson et al., 2002). Microbially enhanced precipitation under oxidizing conditions might explain the late formation of iron oxides/-hydroxides and manganese oxides (cf. Chan et al., 2000; Sedat et al., 1992), which may block pore throats, reducing the permeability of the sandstones in both locations (see Fig. 2.14). However, these late uplift authigenic phases are restricted to sandstones in outcrop analogs and can therefore be neglected when predicting subsurface reservoir characteristics.

The major increase in permeability by one order of magnitude from the lower to the upper Westphalian D sandstones is due to differences in the type of clay cement present. These differences depended on the climatic conditions during and shortly after deposition, and were potentially enhanced by variable pH conditions due to the occurrence of the copper shales at the Hüggel location (Fig. 2.19). However, the low correlation coefficients (see Fig. 2.14) suggest that these processes had a minor impact on reservoir quality within a particular location. In the sandstones studied, late uplift-related cements overprinted the reservoir characteristics.

Thus, the identification of sandstones with a low detrital feldspar content which were deposited under arid climate conditions during the Upper Carboniferous might be incorporated into future exploration strategies if the diagenetic alterations, i.e. clay mineral authigenesis, and burial conditions can be further constrained.

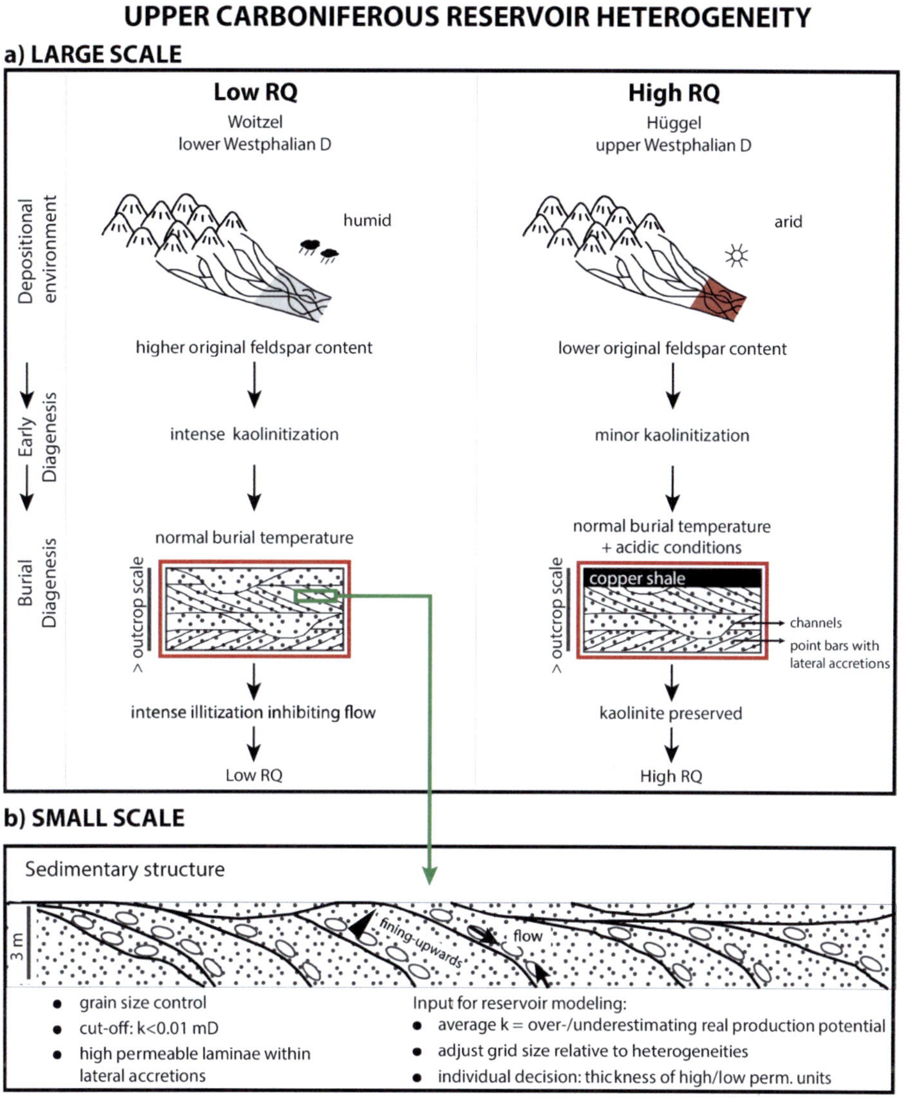

Figure 2.19. Summary diagram showing large (stratigraphic)- and small (outcrop)-scale heterogeneities in the lower and upper Westphalian D sandstones studied at the Woitzel and Hüggel locations. a) Variations in the depositional environment and climate during the early and late Westphalian D controlled diagenetic processes. Humid climate conditions during early Westphalian D deposition and high detrital feldspar contents favor intense kaolinitization. This kaolinite can be replaced by illite during deeper burial resulting in low reservoir qualities of the Woitzel sandstones. By contrast, arid climatic conditions and the later effects of an acidic environment preserved kaolinite and result in better reservoir quality of the upper Westphalian D sandstones at Hüggel. b) Schematic section of the quarry wall at the Woitzel location (redrawn from Fig. 2.16). Lateral variations in reservoir quality of the point bar sandstones are due to the geometry of the depositional units, and to fining-upwards grain-size variations and lateral accretion surfaces (black lines). These small-scale heterogeneities should be taken into account during reservoir modeling.

2.7 Conclusions

The study of two tight Upper Carboniferous sandstone reservoir analogs shows that the reservoir characteristics improve from the early Westphalian D Woitzel sandstones to the late Westphalian D Hüggel sandstones. The permeability varies by one order of magnitude from 0.03 mD (Woitzel) to 0.54 mD (Hüggel) across a vertical stratigraphic distance of approximately 600 m. Laterally, the permeability in a single sandstone bed changes by two orders of magnitude over a distance of 150 m. Different controlling factors have been evaluated at different scales.

Differences in diagenetic minerals control the large-scale permeability variations. The lower reservoir quality of the lower Westphalian D is due to the higher authigenic illite content (up to 12%), while the main clay phase of the upper Westphalian D is kaolinite (up to 11 %). The amount of authigenic quartz cement, approximately 3%, is similar in samples from both study areas. The porosity is mainly secondary and is related to the dissolution of authigenic carbonate cements and aluminosilicates. The large amount of microporosity related to the formation of authigenic clay minerals contributes to the secondary porosity.

The higher original feldspar contents (4.7%) and more humid depositional conditions of the lower Westphalian D sandstones, relative to the lower feldspar contents (1.5%) and the arid conditions of the upper Westphalian D sandstones, favored early kaolinitization and subsequent illitization.

At an outcrop scale, the upper Westphalian D tight sandstones exhibit a positive trend between grain size and permeability. The lower Westphalian D sandstones show a similar trend but only if the data set is limited to samples with permeability values above 0.01 mD.

3 Outcrop to well – Integrating the reservoir quality evolution of Upper Carboniferous tight gas sandstone analogs and reservoirs, NW Germany

3.1 Abstract

A better understanding of reservoir quality variations in tight, gas-producing sandstones is key to develop successful exploration strategies. This industry-academia collaboration focuses on the integration of data obtained from three outcrops (Piesberg, Woitzel, and Hüggel) and two gas fields (drilled by Well A and Well B) of the Upper Carboniferous (Westphalian C/D) of the Lower Saxony Basin, NW Germany. A combination of petrophysical data sets and the diagenetic evolution associated to the burial- and structural history of the basin is used to delineate controlling factors on reservoir quality on a basin-scale.

Similar matrix permeabilities occur in the samples of two outcrops (Woitzel, Hüggel) and the subsurface sandstones, but porosities vary. Permeability reduction during burial is related to clay cement formation mainly controlled by variations in the original sandstone composition of the depositional environment and climate. Differences in the burial history due to basin inversion tectonics result in higher matrix porosities in the uplifted reservoir of field B compared to field A by reducing the formation potential of quartz cements. Further increased porosities in the outcrop sandstones are related to dissolution processes of carbonate cements in the late stages of uplift. Temperature anomalies in the vicinity of faults significantly deteriorate reservoir quality by enabling intense additional mineral cementation as shown in the Piesberg sandstones. In contrast, faulting or natural fracturing may enhance reservoir quality within one field by acting as potential migration pathways for leaching fluids or fracture permeability with production potential.

This complex interaction of the depositional setting, burial related diagenetic processes, and structural characteristics needs to be carefully evaluated. An improved understanding of those processes is the foundation for future exploration and appraisal efforts in the tight gas sandstones of the Lower Saxony Basin and comparable areas.

3.2 Introduction

Upper Carboniferous tight gas-producing sandstone reservoirs have been known to bear great exploration potential in NW Europe for the last decades (e.g., Besly, 1998; Jones and Glover, 2005; Kombrink et al., 2010; Leeder and Hardman, 1990). However, various uncertainties still impede continuous successful exploration attempts. One of the main reasons for this is challenging underground mapping due to great overburden of up to 3.5 km thickness containing thick Late Permian Zechstein evaporite sequences (up to several hundreds of meters, e.g., Peryt et al., 2010) that strongly reduce the quality of seismic imaging (Besly, 1998). Another reason is related to primary depositional heterogeneities in the Upper Carboniferous fluvial successions (e.g., Kombrink et al., 2010). Fluvial sandstones generally show lateral variations in depositional facies and channel geometry variations (cf. Miall, 1988; Pranter et al., 2007), which complicate well planning (Besly, 1998). In addition, smaller scales of heterogeneities occur, from internal grain-size trends (Miall, 1988; Pranter et al., 2007) to complex diagenetic mineral formation and dissolution processes.

The Upper Carboniferous fluvial sandstones often appear to be heavily cemented (Besly, 1998; Lüders et al., 2010) explaining why they are generally referred to as 'tight' (permeability <0.1 mD for US standards, Holditch (2006); <0.6 mD for North German standards, (Albrecht and Reitenbach, 2015; Pusch et al., 2005)). Thus, structurally influenced areas, e.g. inverted blocks, are favored exploration targets (e.g., Hollmann et al., 1997; Kombrink et al., 2010; Lüders et al., 2010) following the general assumption of finding higher fracture intensities around faults and thus, fluid migration pathways despite the low matrix permeability (cf. Caine et al., 1996). However, significant exploration and development risks are attributed to the presence of thermal anomalies, which have been studied in detail for the German area of known Upper Carboniferous tight gas fields (e.g., Cameron and Ziegler, 1997; Lüders et al., 2012; Petmecky et al., 1999; Senglaub et al., 2005a; Senglaub et al., 2005b; Teichmüller, 1986). Wüstefeld et al. (2017b) show that such high temperature anomalies can be related to hydrothermal fluid circulation along fault zones resulting in kilometer wide zones of significantly reduced reservoir quality.

The use of outcrop analogs for fluvial settings in an attempt to improve reservoir characterization by adding information about lateral variations is widely applied (cf. McKinley et al., 2011; Miall, 1988; Pranter et al., 2007; Stephen and Dalrymple, 2002; Trendell et al., 2012). For the Upper Carboniferous reservoirs, exposed rocks that could potentially serve as analogs are generally scarce due to the thick post-Carboniferous overburden (cf. Besly, 1998; Kombrink et al., 2010). However, early works of UK outcrops introduced stratigraphic synthesis of great importance for the Lower Carboniferous onshore UK fields (Besly, 1998). As coal-bearing and red-bed Upper Carboniferous sequences are poorly exposed in NW Europe, the rare outcrops along the southern margin of the Lower Saxony Basin, NW Germany, are regarded as potential analogs not only for the NW German gas fields, but also for the fields onshore the UK, the Netherlands, and the North Sea (cf. Besly, 1998; Jones and Glover, 2005; Kombrink et al., 2010) (Fig. 3.1a). Thus, several studies on the NW German Upper Carboniferous outcrops have been published in the last years with primary focus on the sedimentology and mineralogy of the sandstones, fluvial depositional geometries, and cyclic sequences (David, 1990; Jones and Glover, 2005; Sedat et al., 1992). More recently, petrophysical data have been added and linked to structural and diagenetic characteristics to explain lateral and vertical porosity and permeability variations (Becker et al., 2017a; Wüstefeld et al., 2017a; Wüstefeld et al., 2017b).

Based on those results of a regional understanding of reservoir heterogeneities and their underlying causes, we aim at linking the data obtained from three Upper Carboniferous outcrops (Piesberg, Woitzel, and Hüggel) to actual subsurface data of two gas fields (Field A and B, drilled by Well A and Well B) (Fig. 3.1b and c). While former studies dealt with heterogeneities within one studied area (Wüstefeld 2017a, b; Becker 2017) and the comparison of nearby areas (Woitzel and Hüggel outcrops, see Becker et al. 2017a; Well A and B in prep.), this study aims at the integration of the different data sets over basin-wide distances. With a focus on matrix porosity and permeability, we compare the contribution of diagenesis on reservoir quality through time and evaluate the effect of structural characteristics of the Upper Carboniferous sandstones to explain differences in reservoir characteristics of the outcrop and subsurface data. By integrating sedimentological, stratigraphic, diagenetic, and structural data of different scales, we propose an exploration strategy for the Upper Carboniferous tight gas sandstone reservoirs in the Lower Saxony Basin.

3.3 Geological setting

The investigated outcrops represent the only three locations of exposed Upper Carboniferous strata in NW Germany. They are located on the SW border of the Lower Saxony Basin at a distance of approximately 50 km to the Upper Carboniferous tight gas fields of this study in the north of this basin (Fig. 3.1b and c). The Lower Saxony Basin formed as part of the Central European Basin System due to Early Mesozoic rift and wrench tectonics (Betz et al., 1987; Gast and Gundlach, 2006; Van Wees et al., 2000; Ziegler, 1990).

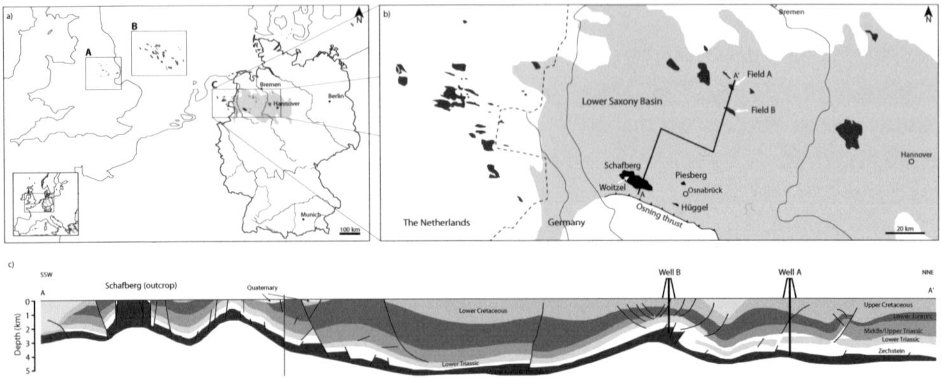

Figure 3.1. a) Location map showing Upper Carboniferous hydrocarbon fields (black) in NW Europe and the Lower Saxony Basin outline (gray shading), A = onshore UK, B = North Sea, C = onshore Netherlands and Germany (modified after Kombrink et al., 2010). b) Close-up of C with the locations of the Upper Carboniferous outcrops at the Piesberg, Schafberg, and Hüggel (black) and of the studied tight gas fields field A and field B (subsurface hydrocarbon fields in dark gray, dashed line = border between the Netherlands and Germany) (modified after Drozdzewski, 1985; Kombrink et al., 2010; Senglaub et al., 2005b). c) SSW – NNE profile through the Lower Saxony Basin highlighting the structural positions of the outcrops in the south and the studied reservoirs in the north of the basin (modified after Baldschuhn et al., 2001).

The Upper Carboniferous sandstone reservoirs in NW Germany have been successfully produced for the last decades, but still hold great potential as reservoir settings are structurally and diagenetically complex (Jones and Glover, 2005). In general, fluvial sandstones feature the most important reservoir units within the Upper Carboniferous in the Central European Basin System (Kombrink et al., 2010). Fluvial depositional systems were mainly present since the Namurian C within the Lower Saxony Basin resulting from marine regression (David, 1990; Sedat et al., 1992). The Westphalian C and D units reach a total thickness of 1500 m in this basin (David, 1990) consisting of cyclic successions that resulted from the waxing and waning of the Gondwana ice shield (Rygel et al., 2008). During the Westphalian C and D, increasing sand proportions from around 60% (Westphalian C, Drozdzewski, 2005) to up to 80% (Westphalian D, David, 1987; David, 1990) indicate a general coarsening-upwards trend. This change in deposition is accompanied by changing climatic conditions towards increasingly arid conditions in the late Westphalian D (Besly et al., 1993; David, 1990; Sedat et al., 1992). As a result, red-colored sand- and siltstones were deposited instead of floodplain sediments and coal seams as recorded from the Lower Saxony Basin (David, 1990), the onshore Dutch fields (Kombrink et al., 2007), and the southern North Sea fields (Besly et al., 1993; Besly and Fielding, 1989; Leeder and Hardman, 1990; Pearce et al., 2005).

Westphalian coals are known to be the main source for the natural gas accumulations in the Upper Carboniferous fluvial sandstone reservoirs (Kombrink et al., 2010), and also for the Permian Rotliegend sandstone and Zechstein carbonate reservoirs in the Central European Basin System (Glennie, 1986; Kombrink et al., 2010; Leeder and Hardman, 1990; Petmecky et al., 1999).

Late Permian Zechstein evaporites act as a basin-wide seal for all those reservoirs (Kombrink et al., 2010; Petmecky et al., 1999). Additionally, the coal-bearing Upper Carboniferous itself acts as a stacked reservoir due to numerous shale horizons at the top of each cycle. Structural traps are generally bound to pop-up structures or anticlines associated with former Variscan wrench-fault systems and controlled by Late Cretaceous inversion (Kombrink et al., 2010). The studied Upper Carboniferous outcrops also developed from such settings. The Piesberg quarry represents an anticline related to the Variscan Piesberg-Pyrmont axis (Drozdzewski, 1985; Drozdzewski et al., 2009), and the Woitzel quarries, belonging to the Schafberg block, and Hüggel complex are interpreted to have developed as pop-up structures related to former Variscan faults (Drozdzewski, 1985; Haglauer-Ruppel, 1991). The major phase of uplift of those structures is controlled by Late Cretaceous inversion tectonics associated with the Africa-Iberia-Europe convergence (Betz et al., 1987; Kley and Voigt, 2008; Kombrink et al., 2010; Lohr et al., 2007; Pharaoh et al., 2010; Senglaub et al., 2005b). This resulted in subsequent inversion on the hanging wall of the Osning thrust (Fig. 3.1b) resulted in the present surface exposure (Baldschuhn and Kockel, 1999; Drozdzewski, 1985). Likewise, several fields in the Netherlands (Kombrink et al., 2010) and NW Germany (Hollmann et al., 1997), as e.g. the studied field B (Fig. 3.1c) (Baldschuhn et al., 2001), indicate such former Variscan structures that were uplifted by Late Cretaceous inversion tectonics. In total, the studied outcrops were uplifted by up to 5 km (Bruns et al., 2013). In contrast, field A underwent an uplift of approximately 1 km up to its present depth of 4 km and field B is nowadays situated at a depth of 2.5 km indicating a stronger uplift of approximately 2.5 km (Fig. 3.1c).

3.4 Materials and methods

Reservoir heterogeneities were studied on different scales using optical microscopy analyses and petrophysical measurements. For that purpose, three outcrops of the Piesberg, Woitzel, and Hüggel and data of two fields (Well A and B) were investigated.

Outcrop rock sampling covers vertical and horizontal profiles (cf. Becker et al., 2017a; Wüstefeld et al., 2017a) with sandstone samples drilled horizontally in the quarries using a water-cooled core dill. Lithological descriptions of the outcrops were supplemented by core descriptions of Well A (approximately 120 m) and B (approximately 308 m).

For petrophysical measurements, 84 outcrop samples (Piesberg n=36, Woitzel n=33, and Hüggel n=15) taken as in-situ drilled plugs with standardized 2.54 cm (1 inch) diameter and maximal 5 cm length were used. Prior to measuring permeability and porosity, all samples were oven-dried at 60°C for a minimum of 10 days. Additionally, 86 porosity and permeability reservoir values (Well A n=43, Well B n=43) were provided from the two different fields for comparison with the outcrop analog results.

Permeability was measured using single-phase gas flow in an isostatic flow cell on those outcrop plug samples (cf. Becker et al., 2017a; Wüstefeld et al., 2017a). Helium (η=1.98*10^{-5} Pa*s) was used as permeant with pressures in the range of 1.5 to 9 MPa according to sample properties. Water was used to build up confining pressures of up to 30 MPa. The permeability values were determined with Darcy's law on the relation between the flow rate Q (m/s), the cross section area A (m^2) of the plug sample, and the pressure gradient dP (P_1 – P_2) (in Pa). With k (m^2) being the permeability coefficient, η (Pa*s) being the viscosity and l (m) being the length of the plug sample, the permeability can be determined as follows (e.g., Amann-Hildenbrand et al., 2016):

$$k = \frac{dV * 2 * \eta * P_2 * l}{dt * A(P_1^2 - P_2^2)}$$

In contrast to outcrop permeability values at confining pressures, provided reservoir data were determined at 20 bar (Well B) or 30 bar (Well A) confining pressure, using dry, oil-free lab-air as the permeant. For linkage of outcrop and reservoir data, understanding the stress dependency of permeability is essential (cf. David et al., 1994; Shanley et al., 2004). To analyze the effect of compaction-related permeability variations, Woitzel, Hüggel, and Well A samples were additionally measured at different confining pressures. David et al. (1994) describe the pressure sensitivity with the following equation:

$$K = K_0 \exp(-\gamma * (P_{eff} - P_0))$$

with K_0 being the determined permeability (mD), P_0 (here: 2 MPa) the reference pressure, and P_{eff} the effective pressure (outcrop samples: 10 or 30 MPa, Well A: 3 MPa). The exponential coefficient is the pressure sensitivity coefficient γ (MPa^{-1}) describing permeability (mD) as a function of the effective confining pressure (MPa) (David et al., 1994). In this study, all outcrop samples and Well A samples were corrected for 2 MPa confining pressure, which allows to directly compare all of them.

Porosity measurements on the outcrop plug samples was performed using a semi-automated gas displacement pycnometer AccuPyc II 1340 by micromeritics®.

Petrophysical results were complemented by petrographic analyses to link diagenetic mineral assemblages to reservoir quality variations. 43 thin-sections of Well A and 51 thin-sections of Well B were provided by the operator to compare the reservoir mineral assemblages to outcrop thin-section results (Woitzel: 29, Hüggel: 11, Piesberg: 51). All thin-sections were polished and stained with yellow-dyed, fluorescent, or blue-dyed epoxy-resin to visualize open pore space. Samples from well A and B were stained by Alizarin Red S to aid identification of carbonate minerals. Samples from the Hüggel outcrops, containing carbonate cements, were stained with a mixture of Alizarin Red S and potassium ferricyanide to discriminate iron-rich carbonate species. A Leica DMLP transmitted-light microscope fitted with a Jenoptik Progres Gryphax Subra digital camera was used for optical analyses. For quantification of the diagenetic mineral assemblage, thin-sections were point-counted by a minimum of 300 grains using a Pelcon Automatic Point Counter mounted to a Zeiss Axiolab A1 or Leitz Aristomet. Grain sizes were determined by measuring the long-axis of at least 100 grains on a grid adjusted to the maximum grain size (e.g., Busch et al., 2017).

To quantify the effect of thermal exposure on the studied sandstone mineralogy, the thermal indicator calculation is exemplarily presented for these lithologies. The thermal indicator in this study is the sum of temperatures at 1 Ma intervals normalized to the total number of Ma since deposition of the lithologies. The burial history after Bruns et al. (2013) is used for this calculation with paths for base Westphalian D used for the Piesberg and Woitzel outcrops and the two fields, and the top Westphalian D path is used to approximate the temperature evolution of the Hüggel sandstones. For the Piesberg sandstones, that were subject to a hydrothermal event, temperatures were adjusted to maximum temperatures related to quartz phases (250°C at around 160 Ma, and 180°C at 90-100 Ma) as shown by Wüstefeld et al. (2017b). The index calculation is given as:

$$Thermal\ indicator = \frac{\sum_{n=depositional\ age}^{today} T_n}{n},$$

where n is the depositional age (in Ma) and Tn is the temperature at a given 1 Ma time step during the burial history.

Applying the above-mentioned methods allows to depict reservoir heterogeneities on an intra-field scale (µm to 100's of meters). Comparing the results of the different outcrop and reservoir locations potentially depicts larger, inter-field and basin-scale (approximately 50 km) variations. Reservoir quality results were linked to their diagenetic characteristics associated to variable burial histories and structural characteristics of the Lower Saxony Basin.

3.5 Results

3.5.1 Lithology

The sandstones of this study comprise of Upper Carboniferous Westphalian C to D strata, which were deposited in a fluvial depositional setting.

In the Piesberg quarry, a 135 m succession of upper Westphalian C to Westphalian D strata was described by Wüstefeld et al. (2017a). Fining-upwards cyclic deposits from conglomerates to sandstones grade into siltstones and coal seams. Sandstone units can feature thicknesses of up to 40 m and they are built up of compound channel sandstones reaching lateral extents of up to tens of kilometers (cf. Jones and Glover, 2005; Wüstefeld et al., 2017a). While downstream accretionary features can be observed in the lower parts of a fluvial cycle, the upper parts are dominated by lateral accretion surfaces (cf. David, 1990; Jones and Glover, 2005; Wüstefeld et al., 2017a). In total, six coal seams are exposed that can be regionally traced over the basin (cf. Klassen, 1984) (Fig. 3.2a). The Zweibänke coal seam marks the border between Westphalian C and D deposits and the Bänkchen coal seam can be correlated to the Alexander coal seam that crops out in the Woitzel quarries, which helps correlating the outcrops (cf. Klassen, 1984) (Fig. 3.2a and c).

In the Woitzel quarries, at a distance of approximately 20 km to the Piesberg, 37 m of lower Westphalian D fluvial deposits showing fining-upwards cycles from conglomerates or coarse-grained sandstones to siltstones and coal seams have been reported (Becker et al., 2017a). The main sandstone unit has a thickness of up to 15 m and consists of single- to multi-storey bar sandstones. Similar to observations in the Piesberg quarry, depending on the position within the quarry, downstream (lower part, start fluvial cycle) or lateral (upper part, later cycle deposition, passes into siltstones) accretion surfaces can be observed. Large channel sandstones also occur in this quarry (Fig. 3.2c) (for detailed description see Becker et al., 2017a).

In the Hüggel outcrops, which are located at a distance of approximately 10 km to the Piesberg and 20 km to the Woitzel quarries, a maximum of 5 m of upper Westphalian D sandstones are exposed (Fig. 3.3). Due to intense weathering, sedimentary features are difficult to distinguish. In general, gray colored sandstones are unconformably overlain by Upper Permian (Zechstein) copper shale and carbonates (Fig. 3.2b). With increasing distance to the unconformity, sandstones show yellow to red colors (for detailed description see Becker et al., 2017a).

Figure 3.2. Overview photographs of the three studied Upper Carboniferous outcrops. a) The Piesberg quarry with the exposed coal seams that allow regional correlation of the stratigraphy. b) The Hüggel outcrop showing the Carboniferous/Permian unconformity, where Late Permian (Zechstein) copper shale transgressively overlies Upper Carboniferous sandstones and conglomerates. c) The Woitzel quarry with the Alexander coal seam that can be directly correlated to the Bänkchen coal seam of the Piesberg quarry (cf. Klassen, 1984). A large fluvial channel can be observed in the quarry wall; dashed line = obscured presence on the S – WSW striking wall section.

The studied Well A is located at a distance of approximately 50 km to the Piesberg. A total length of 479 m of Westphalian C and D are drilled with 120 m of core material (Fig. 3.3). Core and logging data show fluvial fining-upwards cycles dominated by sandstones that grade into siltstones and coals. Similar to the outcrop observations, the sandstone deposits at the base of those cycles are characterized by downstream accretion with a change towards lateral accretion surfaces towards the upper parts (cf. Field X, Wimmers and

Koehrer, 2014). In general, Westphalian C sandstones occur mainly coarse-grained and the present silty intervals show gray to black colors and rootlets. Those units are intercalated with coal seams. Within the Westphalian D, sandstones show finer grain sizes towards the top, where silty intervals show red colors and lack coal deposits.

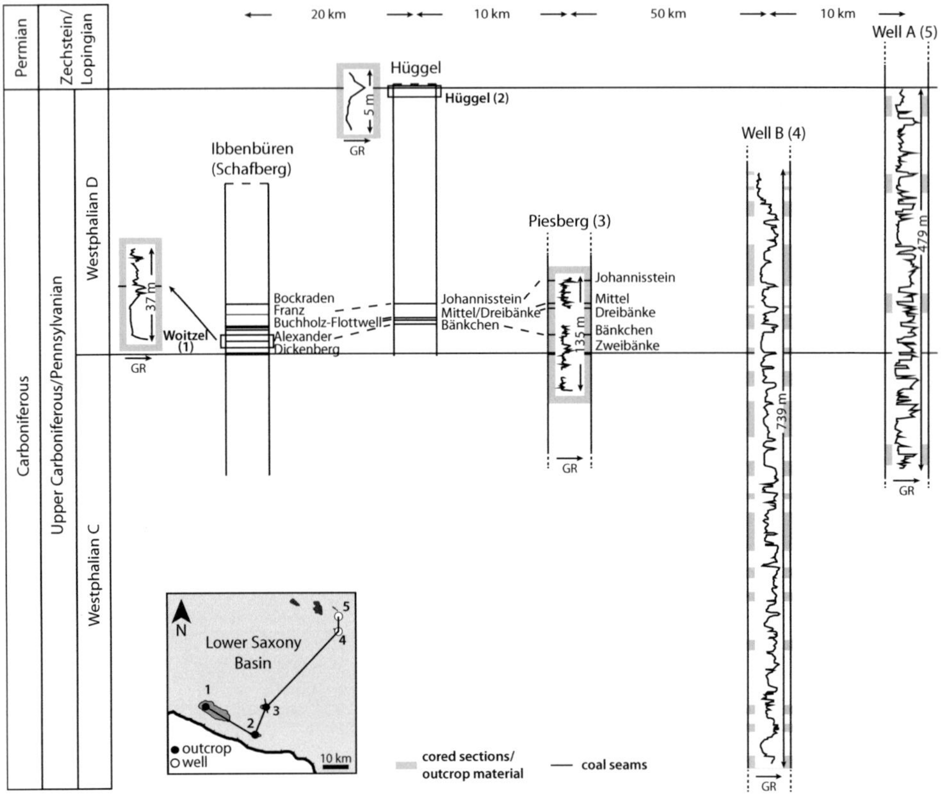

Figure 3.3. Correlation of the studied outcrops (1,2,3) and wells (4,5). Coal seams can be directly correlated between the different outcrops (modified after Klassen, 1984), and the Westphalian C/D boundary can be traced due to the occurrence of the Dickenberg or Zweibänke coal seam in the outcrops (Klassen, 1984) and due to biomarker analyses in the wells.

Well B is at a distance of approximately 10 km to Well A and 40 km to the Piesberg. Here, 739 m of Westphalian C and D strata were drilled and 308 m of core material was taken (Fig. 3.3). In similitude to results of Well A, core and logging data support the observations of fluvial cyclic deposition. In contrast to Well A, the main drilled part contains Westphalian C deposits featuring generally medium-grained sandstones, black silt-rich intervals, and intercalated coal seams. In Well B, the Westphalian D rocks show similar characteristics as the Westphalian C. While the sandstones of Well A are mainly tectonically undisturbed with a total of one open and four partly cemented fractures reported for the entire cored interval, Well B sandstones are intensely fractured with 110 open and 51 partly cemented fractures in the cored interval.

3.5.2 Petrography

Understanding the mineral composition and the distribution of authigenic cement phases is essential to analyze the reservoir quality and compare differences between the studied sandstone samples of the different locations. Point-counting on 91 outcrop- and 94 subsurface sandstone samples shows that all studied Upper Carboniferous sandstones are of lithic to sublith-arenitic composition with generally low feldspar contents following the classification after McBride (1963) (Fig. 3.4, Table A.1). All sandstones show wide ranges of grain sizes (Piesberg: fine- to coarse-grained, cf. (Wüstefeld et al., 2017a); Woitzel: medium- to coarse-grained, Hüggel: fine- to coarse-grained, cf. (Becker et al., 2017a); Well A: fine- to medium-grained, Well B: very fine- to coarse-grained), but with average grain sizes of 0.4 mm in the outcrop sandstones and 0.3 mm in the subsurface sandstones. All sandstones appear moderately to moderately well sorted (Table A.1).

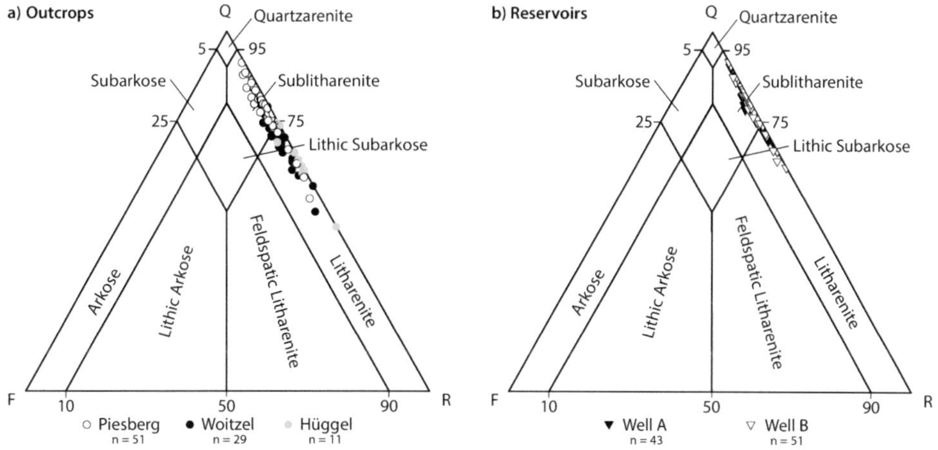

Figure 3.4. Classification of the studied Upper Carboniferous outcrop (a) and reservoir (b) sandstones on a QFR triangular plot (plot after McBride, 1963), all showing lithic to sublith-arenitic compositions. Present-day feldspar contents as derived from point counting are shown in both plots. a) Woitzel (black) and Hüggel (gray) sandstone samples (Becker et al., 2017a) contain slightly higher amounts of lithic components compared to Piesberg (white) sandstones (Wüstefeld et al., 2017a). b) Sandstone compositions of Well A (black triangle) and Well B (white triangle) appear to be comparable to the outcrop sandstone compositions.

For the purpose of this study, point-counting categories are summarized to focus on the main differences between the studied locations rather than on giving detailed descriptions on the mineral grain assemblage. Therefore, detrital components only comprise of quartz, feldspars, and rock fragments. Pseudomatrix, which is widely used to describe highly deformed ductile lithic grains and often occurs in fluvial sandstones (cf. Füchtbauer, 1988; Leeder and Hardman, 1990; Worden and Burley, 2003), is thus assigned to the rock fragment category. Figure 3.4 shows that all studied sandstone samples are well comparable in their detrital compositions with similar amounts of quartz, feldspar, and rock fragments. Mica, matrix, and mineral grains and accessories are further distinguished as detrital components but show similarly low amounts in all studied samples (Table A.1).

With regard to authigenic cement phases, cement neomorphisms and replacements are generally distinguished. Quartz cement is present in all studied sandstone samples (Figs. 3.5a and 3.6a,b, and d). While significant amounts occur in the Piesberg sandstone samples (average 12.7%, max. 25%, min. 1.3%) and Well A samples (average 12.4%, max. 20%, min. 4.3), the amount of quartz cements appears to be clearly lower in the sandstone samples of Well B (average 6.6%, max. 12.0%, min. 1.3%). For the Woitzel (average 3.7%, max. 7.7%, min. 0%) and Hüggel (average 2.6%, max. 5.3%, min. 0.3%) sandstones, the amounts of quartz cement are even lower.

In contrast, the amount of total clay cements, consisting of mainly illite and kaolinite, and partially chlorite (for Piesberg, Well A and B), appears to be more evenly distributed in the studied sandstones (Fig. 3.5b). With an average of 6.8% in Woitzel (max. 12.3%, min. 1.3%) and of 6.6% in Hüggel (max. 11.7%, min. 2.3%) sandstones, those outcrop samples show slightly higher amounts of total clay cements than Piesberg sandstones (average 4.1%, max. 11%, min. 0.3%) and sandstones of Well A (average 3.2%, max. 6%, min. 0%) and Well B (average 3.5%, max. 6.7%, min. 1%).

Carbonate cements are present in the subsurface sandstone samples of Well A (average 6.5%, max. 23.3%, min. 0.7%) and Well B (average 5.8%, max. 18.3%, min. 0.7%) (Fig. 3.6a and b). However, with an exception of four outliers (max. 15.7%, min. 0.3%) of Hüggel sandstone samples and one Piesberg sample (0.7%), carbonate cementation cannot be observed in the studied outcrop sandstones (Fig. 3.5c).

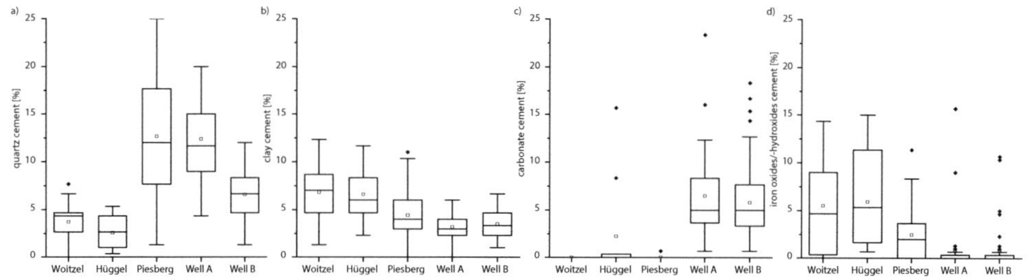

Figure 3.5. Box-Whisker plots of different authigenic cement phases plotted for the different studied locations. a) The amount of authigenic quartz cement appears to be highest in the samples of the Piesberg and Well A with a median of 12.5% and clearly lower in the samples of Well B showing a median of around 7%. Woitzel and Hüggel sandstone samples show even lower amounts of quartz cement. b) Clay cements are more evenly distributed with slightly higher total point-counted amounts in the outcrop than subsurface sandstone samples. c) With the exception of four Hüggel sandstone samples, carbonate cements do not occur in outcrop sandstone samples, but are present to similar amounts both in Well A and Well B sandstone samples. d) Iron oxides/-hydroxides occur frequently in the outcrop sandstone samples but only to a minor amount in the subsurface sandstones. Median = line inside boxes, arithmetic mean = white rectangles in the boxes, lines = 0-100% data range, black rectangles = outliers.

In contrast, iron oxides/-hydroxides occur frequently in the studied outcrop sandstone samples (Fig. 3.5d). With an average of 5.5% in the Woitzel (max. 14.3%, min. 0%) and of 5.9% in the Hüggel (max. 15%, min. 0.7%) samples, there are more iron oxides/-hydroxides observable at those outcrop locations compared to the Piesberg (average 2.7%, max. 11.3%, min. 0%). With the exception of two samples each, showing up to 15.7% (Well A) and 10.7% (Well B) of iron oxides (Fig. 3.6d), those cements are generally present to a minor amount in the subsurface sandstones (Well A and B: average 0.8%).

With regard to clay replacements, illite and/or kaolinite replacing feldspars (Fig. 3.6 and b) or feldspathic rich rock fragments can be observed in all outcrop and subsurface samples except of the Piesberg sandstones. Illite also appears as kaolinite-replacing (Fig. 3.6e) in all samples except those of Woitzel, where illite represents the only clay phase.

In the subsurface samples, some additional cement phases or characteristics occur. Carbonate cements can also appear as replacive cement (Well A: average 0.2%, Well B: average 0.1%) (Fig. 3.6c). As an additional cement phase, barite (Fig. 3.6d) is present in six samples of Well A with up to 6.7% and in four samples of Well B with 0.3%, respectively. In one sample of Well B, pyrite is present with 0.3% (Fig. 3.6e). In two samples of Well B, veins can be observed (Fig. 3.6f).

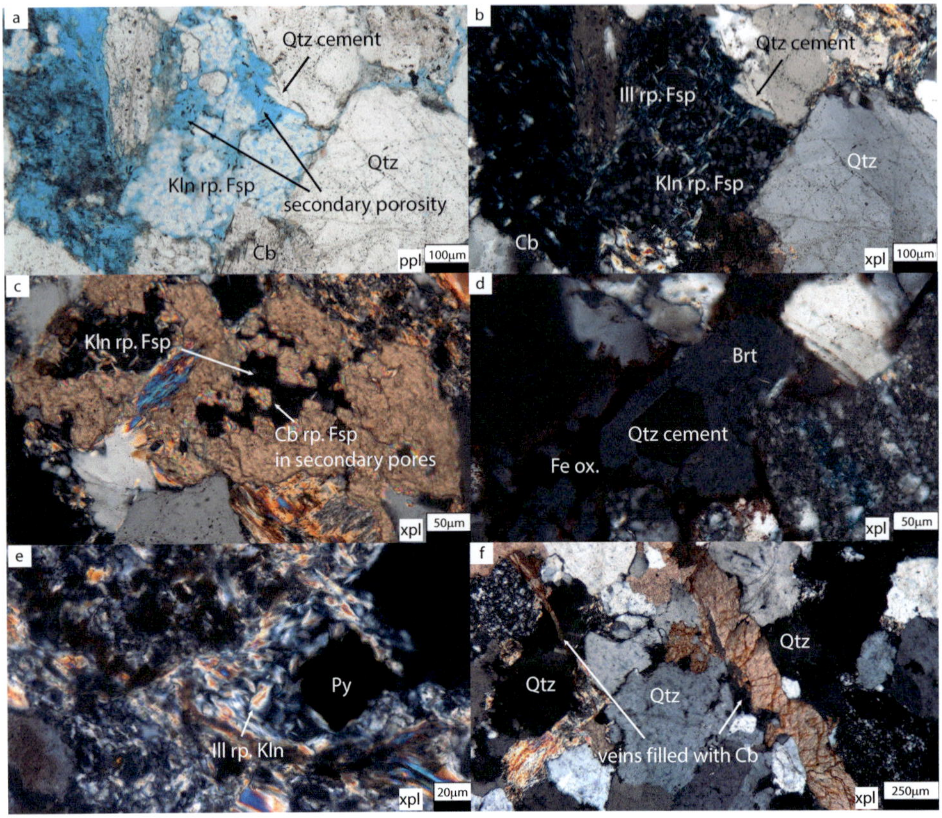

Figure 3.6. Photomicrographs showing replacive cement phases and special features of the Upper Carboniferous sandstone samples. a,b) Quartz cement occuring as syntaxial overgrowth on a quartz grain and carbonate cements as pore-filling. Secondary intragranular porosity generated by detrital feldspar replacement by clay cements, i.e. kaolinite (a and b) and illite (b). Illite appearing to be aligned to the former feldspar cleavage pattern (sample A_391573). c) Carbonate cement within kaolinite replacing a former feldspar grain indicating its formation during grain replacement (sample A_436744). d) Barite cement surrounding an idiomorphic shaped quartz cement and iron oxides filling the pore space (sample A_414568). e) High interference colors of illite on preserved kaolinite booklets indicating the replacement of kaolinite by illite. Pyrite characterized by its opaque appearance and idiomorphic form (sample B_261178). Veins cutting through detrital quartz grains filled with carbonate cements (sample B_273940). Qtz = quartz, Fsp = feldspar, Kln = kaolinite, Cb = carbonate, Ill = illite, Brt = barite, Fe ox. = iron oxides, Py = pyrite, rp. = replacing, ppl = plane polarized light, xpl = cross-polarized light.

3.5.3 Thermal exposure

In an attempt to not only qualitatively link the present mineral cements to their formation time during the burial history but to further quantitatively link the amount of cements to their thermal exposure, thermal indicators were calculated for the studied sandstones.

The determined thermal indicators are highest for the Well A (99 °C/Ma) and Piesberg (97°C/Ma) sandstones. Lower coefficients are calculated for Well B (83°C/Ma) and the Woitzel (70°C/Ma) and Hüggel (64°C/Ma) outcrops.

Those thermal indicators are correlated to the observed amounts of average quartz cementation of each studied area, as temperature is known to be a main controlling factor in quartz cement formation (e.g., Busch et al., 2018; Lander et al., 2008). Different controlling parameters on the syntaxial cementation by quartz (e.g., extensive grain coatings or large variations in grain coating coverage, see Busch et al., 2018) will not be accurately represented by this correlation. With R^2=0.98, results show a strong positive correlation between the amount of quartz cement and thermal exposure of the studied sandstones (Fig. 3.7).

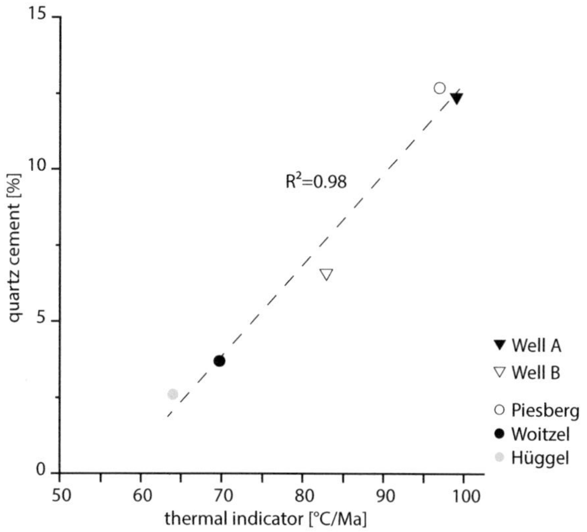

Figure 3.7. Correlation of quartz cement to the calculated thermal indicators of the studied areas showing a positive trend.

3.5.4 Petrophysics

For reservoir quality evaluation all available matrix permeability and porosity data are analyzed. Outcrop data integrates a total of 84 measured porosity and permeability values including 36 Piesberg, 33 Woitzel, and 15 Hüggel samples (cf. Becker et al., 2017a; Wüstefeld et al., 2017a; Wüstefeld et al., 2017b). Provided reservoir data contain petrophysical data of both Well A (total: 43 ϕ and k values) and Well B (total: 43 ϕ and k values) (Fig. 3.8).

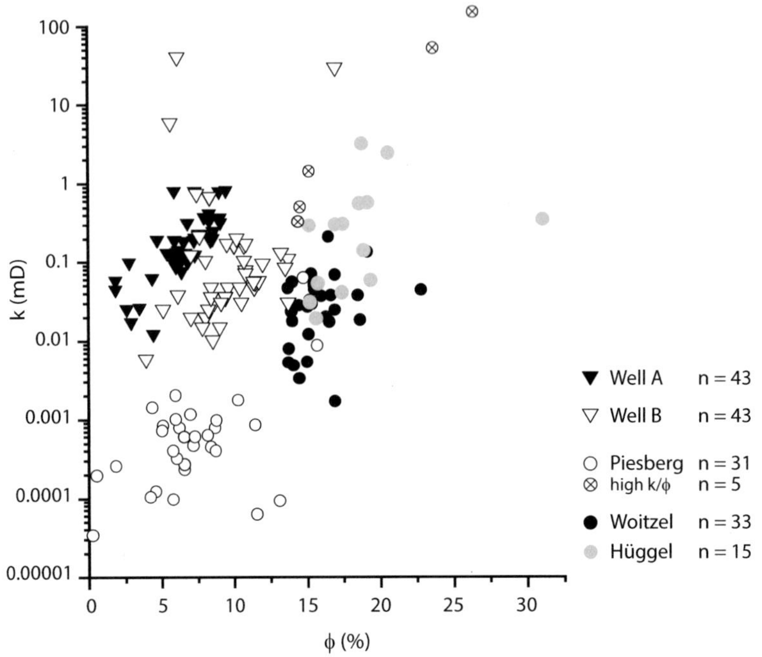

Figure 3.8. Permeability-porosity cross-plot of 84 outcrop sandstone samples (circles) and 86 subsurface sandstone samples (triangles). Measurements on the Piesberg (white circles), Woitzel (black circles), Hüggel (gray circles), Well A (black triangles) samples are corrected for 2 MPa confining pressure. Measurements on the subsurface Well B (white triangles) samples were done at 2 MPa confining pressure. High permeable and high porous Piesberg samples are marked with a cross.

Well A data feature matrix permeabilities in the range of 0.01 to 1 mD and porosities between 2 to 10%, with an average of 6.4% porosity (Fig. 3.8). Well B shows similar ranges of permeability but three outliers towards higher values of up to 100 mD occur. The outliers are related to samples with macroscopically visible fractures. Porosity values of Well B likewise cover a wider range of 4 to 17% porosity resulting in a higher average porosity of 9.6% compared to Well A. Regarding the outcrop samples, Piesberg data feature a similar range of porosities compared to subsurface data (0.2 – 17%) and with 7.4% a similar average porosity. Permeabilities of the Piesberg sandstone samples appear to be lower than subsurface permeabilities with mainly 0.00001 to 0.01 mD. However, five samples are separated showing high porosity (15 - 27%) and permeabilities up to 219 mD. If not explicitly mentioned, the outliers are neglected for the following discussion as they are associated with fault-related leaching processes (Wüstefeld et al., 2017a). In contrast to the majority of the Piesberg sandstones, Woitzel and Hüggel samples feature higher porosities between 15 to 30 % and also higher permeability values in a range comparable to the reservoir data with 0.01 to 10 mD.

3.6 Discussion

Westphalian C and D deposits feature a total thickness of around 1500 m in the Lower Saxony Basin (David, 1990) of which 739 m were drilled by Well B. Together with the observations obtained from Well A (Westphalian C and D), Piesberg (Westphalian C and D), Woitzel (lower Westphalian D), and Hüggel (upper Westphalian D), a wide range of the Westphalian C and D stratigraphy was covered (see Fig. 3.3).

Rock material of Westphalian C and lower Westphalian D mainly represents deposits of braided fluvial systems as indicated by generally coarse grain sizes or conglomeratic composition at the base of fluvial cycles and internal erosive surfaces in the core, and additionally by the observation of downstream accretion surfaces in the Piesberg and Woitzel outcrops. Lower energy deposition of more meandering channel types is characterized by lateral accretion surfaces on bar sandstones in the outcrops, and interpreted by the observation of fining-upwards trends in the individual sandstone units, intercalation of mud-rich intervals towards the top of the sandstone interval, or heterolithic composition if referred to a stacked channel system in the core material. A change in climate from humid to more arid conditions is reported for the ongoing Westphalian D (Besly et al., 1993; David, 1990; Jones and Glover, 2005). As a result, coal seam occurence decreased in the Lower Saxony Basin, and coloration of the sandstones and siliciclastic material changed from gray to red, which is reported throughout NW Europe (Besly et al., 1993; Klassen, 1984; Leeder and Hardman, 1990; Moscariello, 2003). This climate change towards arid conditions is indicated by red sandstones from the Hüggel outcrop (Becker et al., 2017a), and red silty intervals representing soil deposits and few sandstones (e.g. samples A_424348 and A_414568, Table A.1), in the Westphalian D cores of Well A.

In general, outcrop and subsurface Westphalian C and D fluvial sandstones show the same characteristics over the Lower Saxony Basin. The humid intervals of the Westphalian C and lower Westphalian D contain fluvial cycle deposition dominated by high energetic braided systems and intercalated coal seams. The upper Westphalian D units, which deposited under arid conditions, show red colors and lack coal seams.

3.6.1 Variations in paragenesis

Despite similar lithological characteristics, the studied sandstones show variations in their mineralogical composition, which is controlled by the depositional environment, provenance, and climate, but mainly by their diagenetic evolution (cf. Khalifa and Morad, 2015; Kim et al., 2007; Morad et al., 2010).

In terms of detrital composition, all studied outcrop and subsurface sandstones indicate the same range of sublithic to lith-arenitic compositions (see Fig. 3.4). In addition, measured grain sizes are similar (see Table A.1) as comparable intervals were sampled for petrographic and petrophysical analyses. Thus, a control of grain size explaining the porosity and permeability variations between the studied locations can be excluded.

Early burial processes affected all studied sandstones in the same way. Ductile lithic grains and clay-rich matrix, which were deposited in the fluvial settings, are deformed due to mechanical compaction and squeezed between the rigid framework grains reducing primary intergranular porosity and ultimately permeability (Leeder and Hardman, 1990; Molenaar et al., 2015). However, as for a potential grain size control, this process can only explain variations within one studied location due to local facies variations (e.g., Becker et al., 2017a; Wüstefeld et al., 2017a), but not control the general differences in porosity and permeability. Hence, main controlling factors on reservoir heterogeneities between the different study areas are expected to be related to later diagenetic stages.

In all sample locations, quartz cement formation results in a significant reduction of the pore space and thus, deteriorates the reservoir storage potential (e.g., Bjørlykke and Egeberg, 1993; Dutton and Loucks, 2010; Meadows and Beach, 1993; Morad et al., 2010; Zhang et al., 2015). At temperatures exceeding 70°C, it is induced by enhanced pressure solution, which provides the silica for quartz growth in addition to the supply from illite-forming reactions or clay mineral transformations (Bjørlykke and Aagaard, 1992; Bjørlykke and Egeberg, 1993; Chuhan et al., 2001). Sutured grain contacts and the presence of illite (only minor occurrence in the Hüggel samples) represent silica sources in all study areas. However, the amount of quartz cementation varies largely between the different sandstones (see Fig. 3.5a). Highest volumes are reached in the Piesberg sandstone samples (max., 25%, see Fig. 3.5a), which experienced highest temperatures of up to 300°C before peak burial resulting from hydrothermal fluid flow along a large NNW – SSE striking normal fault (Wüstefeld et al., 2017b). For the other study areas, such a thermal overprint is not reported. Thus, the different quartz cement volumes can result from differences in their normal burial histories (Fig. 3.9). Although quartz cement development was shown to be controlled by grain size and available surface area and the volumes formed over time are also changing due to the attaining of euhedral crystal facets (Lander et al., 2008), quartz cement formation also depends on the temperature and is shown to be sensitive to the duration time at such temperatures (Busch et al., 2018; Lander et al., 2008; Makowitz et al., 2006; Walderhaug, 1994, 2000). The results of the thermal indicator calculations indicate this impact of thermal exposure on the development of quartz cement volumes (see Fig. 3.7). Thus, highest quartz cement volumes would be expected for the Well A sandstones that has stayed at its present depth of approximately 4 km since the Late Cretaceous and shows the highest calculated thermal indicator of 99°C/Ma (Fig. 3.9). Slightly less volumes would be expected for the Well B sandstones at a present-day depth of approximately 2.5 km and thus, exposed to lower temperatures than the Well A sandstones resulting in the lower thermal indicator of 83°C/Ma. Lowest volumes should occur in the Woitzel and Hüggel sandstones, that experienced rapid uplift due to Late Cretaceous inversion and final surface exposure in the Quaternary (cf. Bruns et al., 2013; Senglaub et al., 2005a). The additional uplift is represented in their low thermal indicators of 70°C/Ma and 64°C/Ma, respectively. Results of this study match well to that model (Figs. 3.5a, 3.7 and 3.10).

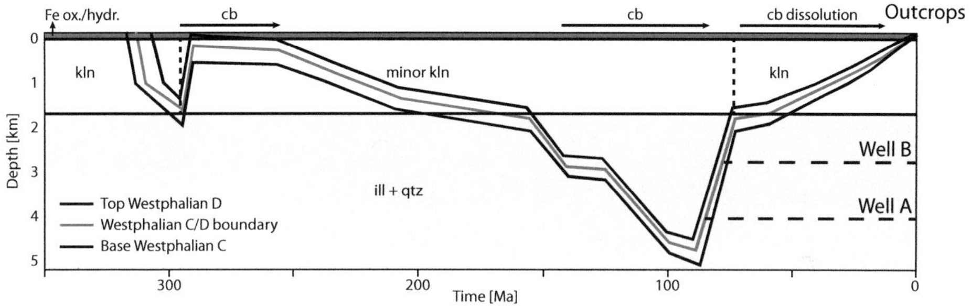

Figure 3.9. Burial history of the Westphalian sandstones (modified after Bruns et al., 2013) with the present location of the studied sandstones. Approximate depth ranges (assumed temperature gradient 30°C/km) of potential formation of the main cement types controlling reservoir quality of the studied sandstones are marked. As the presence of meteoric water is needed for the formation of early kaolinite (e.g., Lanson et al., 2002; Morad et al., 2000), minor formation is expected for times of arid climate conditions. Qtz = quartz, kln = kaolinite, cb = carbonate, ill = illite, Fe ox./hydr. = iron oxides/-hydroxides.

As another process, the formation of clay cement phases, ie. kaolinite, illite, and chlorite, are known to significantly deteriorate the reservoir flow potential (e.g., Bjørlykke and Aagaard, 1992; Bjørlykke and Egeberg, 1993; Chuhan et al., 2001; Cowan, 1993; Gaupp et al., 1993; Gaupp and Okkerman, 2011; Lanson et al., 2002; Morris and Shepperd, 1982; Worden and Burley, 2003; Worden and Morad, 2003). Illite forms if potassium and aluminium are present as e.g. sourced from feldspar and kaolinite (Bjørlykke and Aagaard, 1992; Bjørlykke and Egeberg, 1993; Chuhan et al., 2001; Lanson et al., 2002). Kaolinite formation, in contrast, does not only depend on the presence of feldspar but also humid conditions (cf. Lanson et al., 2002; Morad et al., 2000), and thus, is not restricted to early diagenetic processes but can also occur during uplift to the surface (Fig. 3.9). Becker et al. (2017a) showed illite to be the main cement phase in Woitzel sandstones, and kaolinite in Hüggel sandstones explained by higher original feldspar contents and humid climate conditions in the lower Westphalian D (Fig. 3.10). Piesberg sandstones contain similar amounts of total clay minerals as the other two outcrops (see Fig. 3.5b). Here, high-temperature chlorite (270°C) occurs as additional clay phase due to its hydrothermal overprint (Wüstefeld et al., 2017b). This overprint and the resulting strong quartz cementation filling the pore-filling illite (Fig. 3.10) might result in slightly underrepresented amounts of illite in the samples and explain the lower amount compared to the Woitzel sandstones. Kaolinite and illite are present in the subsurface samples (Fig. 3.10) with lower amounts compared to the outcrop sandstones. The reservoir sandstones were deposited at a more distal location in the basin at a greater distance to the sourcing Variscan front (Kombrink et al., 2010). Together with general NW-ward Westphalian paleoflow directions (Jones and Glover, 2005; Kombrink et al., 2007) this indicates that more feldspar was potentially altered during transportation due to its lower stability as e.g. compared to quartz (Füchtbauer, 1998; Tucker, 2001), and less feldspar deposited in the distal regions. As both, kaolinite and illite, formation processes require a feldspar source, less clay cements are present in the samples due to the potentially lower original feldspar content.

Especially iron-rich carbonate cements are reported for Upper Carboniferous sandstones (Besly et al., 1993; Moscariello, 2003; Sedat et al., 1992), that additionally reduce pore space. Their presence is well documented for the subsurface samples of this study, though they are nearly absent in the outcrop sandstones (see Figs. 3.5c and 3.10). The latter results from dissolution processes by infiltrating meteoric waters during uplift and exposure at the surface (e.g., Ketzer et al., 2003; Lanson et al., 2002). The former presence of those carbonate cements in the outcrop samples is suggested by the observation of rhombohedral shaped secondary, intergranular pores (Fig. 3.10) (cf. Becker et al., 2017a; Wüstefeld et al., 2017a).

Iron oxides/-hydroxides represent another important cement type reducing permeability by blocking pore throats. They mainly formed due to microbial processes under oxidizing conditions (cf. Chan et al., 2000; Sedat et al., 1992), and thus, are associated with late stage uplift near surface conditions (Fig. 3.9). As a result, iron oxides/-hydroxides are present in the studied outcrop samples but almost absent in the subsurface, where only few iron oxide cements are present (see Figs. 3.5d and 3.10). Here, the present cements could have formed due to arid climatic conditions (cf. Worden and Morad, 2003).

Differences in the mineral assemblage of the studied sandstones are mainly associated to later burial diagenetic processes and resulting mineral authigenesis. Main variations are the amount of quartz cement, the type of clay cement, and the presence or absence of carbonate cements or iron oxides/-hydroxides.

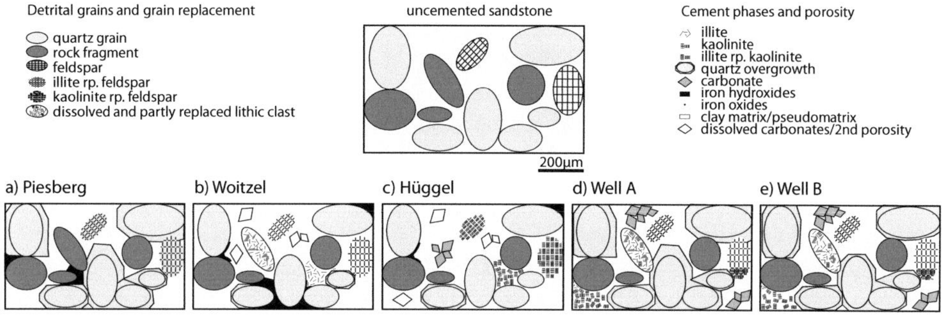

Figure 3.10. Schematic micrographs of the studied sandstones showing main compositional differences between the different study areas. Variations occur in the amount of quartz cement, the type of clay cement, and the presence of iron oxides/-hydroxides. Secondary inter-/intragranular pores and carbonate cements also vary. a) Piesberg sandstones showing large amounts of quartz cement similar to Well A but also iron-hydroxides similar to the other outcrop samples. b) Woitzel sandstones with small amounts of quartz cement and illite as only clay cement phase, compared to c) Hüggel sandstones with kaolinite as main clay mineral cement phase appearing partly illitized. Both sandstone samples show secondary intergranular porosity due to carbonate cement dissolution with few carbonates left in the Hüggel samples. d) Well A sandstones containing large amounts of quartz cement, carbonate cements, and illite and kaolinite cements. e) Well B sandstones appearing similar to Well A sandstones but with less quartz cements. The effect of compaction on grain boundaries is neglected in this sketch.

3.6.2 Reservoir quality control on a basin-wide scale

Porosity and permeability variations occur between the different studied locations. Porosities of the Woitzel and Hüggel outcrops (average 15% and 19%) are higher than those of Well A and B (average 6% and 10%), while Piesberg samples (average 7%) show similar porosities to the subsurface sandstones (see Fig. 3.8). The permeability of the Piesberg sandstones (range 0.00001 to 0.01 mD) are lowest, and the permeability of Well A and B, and the Woitzel and Hüggel outcrops are higher by up to several orders of magnitude (reservoirs: range 0.01 to 100 mD; outcrops: range 0.01 to 10 mD) (see Fig. 3.8, Table 3.1). The latter match well with permeabilities of Westphalian C/D sandstones of the southern North Sea (Leeder and Hardman, 1990; Moscariello, 2003).

Results of this study so far indicated similar stratigraphy and depositional environments, and slightly different detrital rock compositions with regard to the original feldspar content of all studied samples, but reservoir qualities vary between the different study areas as does the composition of authigenic minerals. Thus, linkage of the latter aspects is the key to understand the differences amongst the studied sandstones.

Subsurface Well A and B and outcrop Woitzel and Hüggel data show permeabilities in the same order of magnitude, but porosities are higher in the outcrop samples (see Fig. 3.8).

Hence, all sandstones experienced a strong deterioration in permeability due to the formation of flow inhibiting clay cements (Figs. 3.10 and 3.11, Table 3.1). Porosities are shown to mainly depend on the amount of quartz and carbonate cement. Thus, values are shifted towards higher values in the outcrop sandstones as former carbonate cements are almost completely dissolved due to uplift-related dissolution processes (e.g., Ketzer et al., 2003; Lanson et al., 2002) (Fig. 3.11). In addition, lower volumes of quartz cement grew in the outcrop samples as they were exposed to elevated temperatures for a significantly shorter duration of time. With regard to Well B, porosities are already enhanced compared to values of Well A. In Well B, stronger Late Cretaceous uplift (e.g., Baldschuhn et al., 2001) and corresponding exposure to lower temperatures compared to Well A reduced further quartz cementation. Thus, the amount of uplift and associated quartz cement formation have a significant control on reservoir quality (Fig. 3.11, Table 3.1), which could be quantified and predicted using thermal indicators. Furthermore, Well B sandstones are intensely fractured as a result of the inversion-related reactivation of former Variscan structures (e.g., Hollmann et al., 1997; Kombrink et al., 2010), which adds a significant amount of fracture permeability to the reservoir quality of field B.

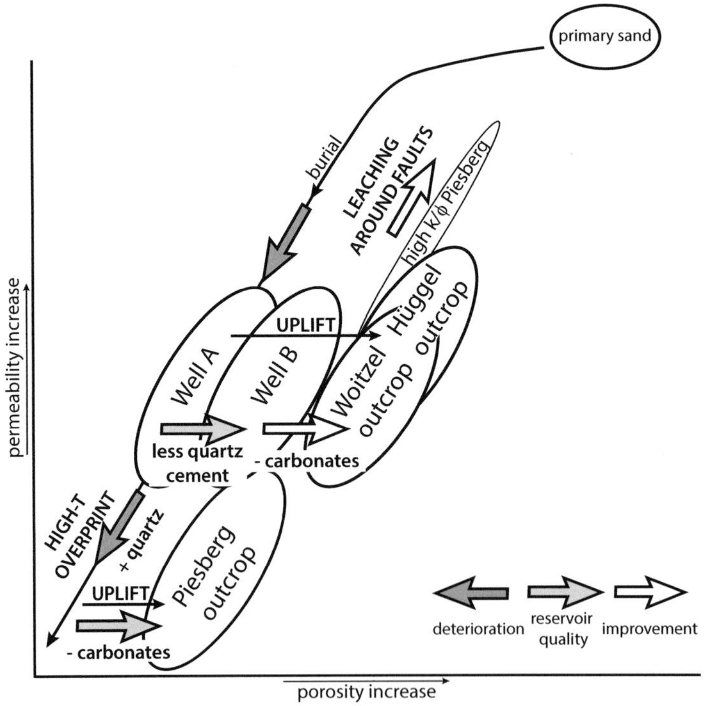

Figure 3.11. Summary sketch of the porosity and permeability evolution during the burial history including controlling factors resulting in reservoir quality variations between the studied locations.

Although exposed at the surface, Piesberg sandstones show low porosities comparable to the subsurface sandstones and even lower permeabilities. Here, a high-temperature overprint resulting from the migration of hydrothermal fluids along a large normal fault caused intense further (quartz) cementation (cf. Wüstefeld et al., 2017b). Thus, if present, a fault-bound thermal anomaly may overprint the normal evolution path and significantly reduce reservoir quality (Fig. 3.11, Table 3.1).

Table 3.1. Summary table and risk assessment of the reservoir characteristics (permeability in mD, porosity in %) and the controlling diagenetic cement phases (in %, derived from point counting) of the studied locations. Red colors indicate poor reservoir qualities and their causes.

Reservoir quality risks		Outcrops			Subsurface	
		Piesberg	Woitzel	Hüggel	Field A	Field B
Average reservoir properties	Porosity (%)	<10	>15	>15	<10	~10
	Matrix permeability (mD)	0.00001-0.01	0.01-0.1	0.1-1	0.1-1	0.1-1
Cementation risks	Quartz cement volume (%)	>10	<5	<5	>10	>5
	Clay cement volume (%)	<5	>5	>5	<5	<5
	Carbonate cement volume (%)	<5	<5	<5	>5	>5
Burial and tectonic history risks	High-Temperature anomaly	Y	N	N	N	N
	Inversion uplift < 2km	N	N	N	Y	N
	Fracture permeabiliy missing	N	Y	Y	Y	N
	Total risk	**Very high**	**Moderate**	**Moderate**	**High**	**Low**

3.6.3 Reservoir quality control on an intra-field scale

Within one study area, variations in reservoir quality are related to primary depositional setting and climate. Generally, finer grain sizes correlate with lower permeabilities in the studied Woitzel and Hüggel sandstones if permeabilities are above 0.01 mD (Becker et al., 2017a). As fluvial sandstones show laterally varying channel geometries and bar sandstones with internal fining-upwards successions, they show large internal variations in permeability (two orders of magnitude over 150 m, Becker et al., 2017a). Furthermore, Wüstefeld et al. (2017a) and Becker et al. (2017a) showed that samples containing a large content of ductile grains, which are deformed during mechanical compaction, feature reduced reservoir qualities (cf. Leeder and Hardman, 1990; Molenaar et al., 2015). For the reservoirs, intervals with lower net-to-gross ratios, describing the percentage of rock volume potentially capable of containing hydrocarbons (Wimmers and Koehrer, 2014), favor smaller grain sizes and higher contents of ductile grains, and thus, lower reservoir qualities.

For areas with normal burial paths and associated temperatures, the order of magnitude of permeability depends on the type of authigenic clay cement present, with generally higher reservoir qualities for kaolinite-dominated sandstones and lower reservoir quality if illite is prevalent (e.g., Gaupp et al., 1993). Becker et al. (2017a) showed that humid depositional conditions and higher original feldspar contents resulted in illite formation in the lower Westphalian D, and that the change towards arid conditions in the upper Westphalian D might indicate a kaolinite dominated setting. Hence, Hüggel sandstones show generally better reservoir qualities compared to Woitzel sandstones (see Figs. 3.7 and 3.10).

Even in the area of the intensely overprinted Piesberg, internal variations in reservoir quality occur (see Fig. 8), that have been studied by Wüstefeld et al. (2017a). W – E striking smaller faults occur within the quarry, and a total of 24 samples was taken within a 30 m distance around those faults (Wüstefeld et al., 2017a). Out of those samples, five samples show unusually high matrix porosity and -permability values (see Fig. 8 and 11), with highest flow values in corridors of up to 5 m around the faults. Wüstefeld et al. (2017a) showed that those samples appear to be strongly leached by circulating high-saline and CO_2-bearing fluids associated to the Late Cretaceous basin inversion (Lüders et al., 2012).

3.6.4 Implications for exploration strategies

This integrated study on the Upper Carboniferous tight gas sandstones of the Lower Saxony Basin indicates complex interaction of the depositional environment of the sandstones, including climatic variations, the burial history with associated temperature evolution and thermal exposure, and the tectonic setting of the field. Those controlling factors induce heterogeneities on different scales, that need to be addressed depending on the modeling purpose.

For the purpose of developing an exploration strategy, basin-scale heterogeneities in reservoir quality need to be taken into account. The largest risk of failure due to intensely deteriorated reservoir qualities is given by the presence of high temperature anomalies. Their presence is well-known for the area of the Lower Saxony Basin, and due to their wide distribution, it was interpreted by various models (e.g., Cameron and Ziegler, 1997; Lüders et al., 2012; Petmecky et al., 1999; Senglaub et al., 2005a; Senglaub et al., 2005b; Teichmüller, 1986). The detailed study on the underlying causes for the thermally overprinted Piesberg sandstones revealed heating of the rocks prior to maximum burial (Wüstefeld et al., 2017b). Hence, conventional reservoir modeling (maximum temperature at maximum burial) (e.g., Bruns et al., 2013) would overestimate the grade of maturation of potential source rocks due to the overestimated burial depth. The largest chance of successful exploration in the tight gas sandstones, in contrast, is associated with normal maximal burial temperatures and following stronger basin inversion. Calculation of the thermal indicators in combination with the amount of uplift is shown to indicate reservoir quality. The comparison of Well B and the Woitzel and Hüggel outcrop to Well A of this study showed that the identification of intervals of high sand net-to-gross ratios for expected large grain sizes in combination with rapid, strong basin inversion are most promising. In addition, deposition under arid conditions reducing the potential of early clay cement formation by feldspar replacement further enhances reservoir quality provided that otherwise equal conditions are present.

For the purpose of modeling one reservoir unit to predict potential flow and storage capacities, internal heterogeneities within fluvial sandstones need to be taken into account. Thus, internal grain size trends related to the fluvial sandstone geometries (e.g., McKinley et al., 2011; Miall, 1988; Pranter et al., 2007) are shown to affect the Upper Carboniferous outcrop sandstones by variations in permeability of two orders of magnitude on a 150 m long single reservoir sandstone1 profile (Becker et al., 2017a). The need for either small or lateral variable grid cell dimensions to correctly estimate permeabilities in detailed models is further addressed by the occurrence of fault-related high permeable zones (5 m distance around faults) (cf. Wüstefeld et al., 2017a).

3.7 Conclusions

This study of three Upper Carboniferous (Westphalian C/D) outcrop analogs and two gas fields at a distance of approximately 50 km highlights their reservoir quality evolution, which is controlled by a complex interplay of depositional environment, burial history, and structural characteristics.

Samples from Well A, Well B, Woitzel, and Hüggel show similar matrix permeabilities (0.01 to 10 mD). Low permeabilites are shown to be generally controlled by diagenetic clay mineral cement formation. Variations in clay mineral cement volumes can be correlated to the depositional content of feldspar minerals.

In contrast, matrix porosities vary between Well A (average 6%), Well B (average 10%), Woitzel (average 15%), and Hüggel (average 19%). Two distinct processes were identified which controlled porosity destruction and enhancement. First, differences in thermal exposures during burial control the quartz cement development. The shift towards better matrix porosities in Well B compared to Well A results from less quartz cement formation. Second, uplift to the surface resulted in additional matrix porosity enhancement in the outcrop sandstones by dissolution of carbonate cements.

Piesberg sandstones show an anomalous development (avg. porosity: 7%, permeability: 0.00001 to 0.01 mD) related to a fault-bound hydrothermal alteration causing intense quartz cementation.

On a basin scale, Upper Carboniferous sandstones in NW- Germany situated in areas of normal burial temperatures and associated to structures that allowed strong uplift during basin inversion represent sweet spots for exploration and production attempts if faults or erosion do not compromise the top seal integrity, while anomalous thermally overprinted sandstones inherit a major exploration risk. The presence of open natural fractures in strongly uplifted areas like field B additionally increases the reservoir quality by the development of dual permeability systems.

4 Correlating three Zechstein Ca2 carbonate outcrops across the Eichsfeld-Altmark Swell – Facies, reservoir properties, and outcrop analog potential

4.1 Abstract

Understanding the lateral distribution of carbonate facies can help improving development strategies as reservoir characteristics vary between different depositional environment types. This study focuses on three outcrops of Zechstein Ca2 (second cycle, Stassfurt) carbonates along the SW margin of the Harz mountains. The carbonates were deposited on the Eichsfeld-Altmark Swell, a NNE – SSW striking paleo-high at the southern margin of the Southern Permian Basin.

For each outcrop, different gross depositional environments (GDE) from slope to platform margin to inner platform deposition are interpreted and linked over a lateral distance of 25 km. Reservoir characteristics are analyzed with regard to the interpreted type of depositional environment and compared to subsurface data of a gas-producing Ca2 field at a distance of approximately 130 km to the NW in the Lower Saxony Basin. Porosity values of the slope carbonates in outcrop are in good accordance with the provided subsurface slope data. Both show a similar range of values and average porosities of approximately 6% (outcrop) and 4% (reservoir). In contrast, permeability values of the subsurface samples are increased compared to outcrop values mainly due to microfracturing. Platform-related grainstone facies show best matrix porosities and –permeabilities and highest proportions of those carbonates are related to the center of the studied Eichsfeld-Altmark Swell. Porosity-permeability relationships are derived for the interpreted gross depositional environments. Together with the presented lateral GDE correlation lengths, they can be incorporated into reservoir models of Ca2 gas fields in NW Germany.

Information about the depositional environment distribution are the base for choosing the correct outcrop analog for future studies of the structural inventory and lateral reservoir quality variations.

4.2 Introduction

Carbonates of the second Zechstein cycle (Ca2, Stassfurt carbonate) contain the largest number of producible hydrocarbon reservoirs within the Southern Permian Basin Zechstein units (Peryt et al., 2010). Reservoir lithologies are mostly located along the southern margin of the basin trending from the Netherlands and Germany to Poland. Successful exploration and production of the Ca2 carbonates has been active for decades (Karnin et al., 1992; Peryt et al., 2010; Strohmenger et al., 1996; Strohmenger and Strauss, 1996). A good reservoir setting is generally present for this unit in this area, as overlying evaporitic successions (A2, Basal anhydrite and younger Zechstein salt sequences) represent a basin-wide top seal for the Ca2 carbonates (Karnin et al., 1996; Peryt et al., 2010; Sannemann et al., 1978; Strohmenger et al., 1996; Strohmenger and Strauss, 1996). Underlying Carboniferous coal measures or organic-rich parts of the Ca2 itself source these reservoirs (Karnin et al., 1996; Peryt et al., 2010; Strohmenger et al., 1996).

A strong diagenetic overprint, containing stages of dolomitization and calcitization, controls the development of the reservoir storage potential and has been in the focus of many studies (e.g., Biehl et al., 2016; Clark, 1980, 1986; Reijers, 2012; Strohmenger et al., 1996; Strohmenger et al., 1998). However, also the occurrence of natural fracture systems influences the reservoir quality of many carbonate reservoirs (e.g., Becker et al., 2018; Gillespie et al., 2011; Lamarche et al., 2012; Wennberg et al., 2006). Furthermore, the distribution of reservoir characteristics has been shown to correspond to the facies distribution of the Ca2 platform along the southern margin of the Southern Permian Basin (e.g., Clark, 1986; Peryt et al., 2010; Reijers, 2012; Strohmenger et al., 1996; Strohmenger et al., 1998).

Most of the previous works are based on subsurface well and seismic data as Zechstein Ca2 outcrops are scarce (Peryt et al., 2010). Problems are therefore associated to limited core material and challenging seismic imaging. The latter is due to the large overburden thickness and overlying salt deposits (cf. Strohmenger et al., 1996). However, the Zechstein Ca2 is exposed along the SW margin of the Harz mountains and their local characteristics have been published in detail in the past (e.g., Herrmann, 1956; Paul, 1991, 1993, 2014). Their deposition is related to the presence of the Eichsfeld-Altmark Swell, a paleo-high containing Permo-Carboniferous plutonic and volcanic rocks (Paul, 1993, 2014) that would be traceable on seismic images. Thus, a correlation of carbonate reservoir characteristics to their position on such a paleo-high would significantly improve development strategies. Becker et al. (2018) recently showed that the exposed fracture characteristics could be incorporated in reservoir modeling of the associated Ca2 reservoirs in the Lower Saxony Basin.

In this study, we aim at a correlation of the known Zechstein Ca2 outcrops along the SW margin of the Harz mountains. Three quarries are investigated in this work. We use a division into three different gross depositional environment (GDE) types to link their characteristics over a lateral distance of approximately 25 km and related to the Eichsfeld-Altmark Swell. We further present petrophysical data associated to those intepreted GDE types and compare them to subsurface data of a reservoir featuring the same stratigraphy approximately 130 km to the NW (Fig. 4.1a) (cf. Field X in Koehrer et al., 2016). Based on the GDE classification, suitable outcrop analogs can be defined for different reservoir facies types.

4.3 Geological setting

The studied quarries are located at the southwestern border of the Harz mountains in North Germany (Fig. 4.1), all exposing Late Permian (Zechstein) Ca2 carbonates. Their deposition took place at the southern margin of the Southern Permian Basin (cf. Pharaoh et al., 2010; Słowakiewicz et al., 2013; Strohmenger et al., 1998), which thermally subsided during the Permian after crustal extension initiated basin formation in the Upper Carboniferous (Scheck-Wenderoth et al., 2008; Van Wees et al., 2000). Transgression of the Zechstein sea during the Late Permian is marked by the basin-wide deposition of the copper shale (Peryt et al., 2010; Pharaoh et al., 2010). In the following 5 to 7 Ma (Menning et al., 2006), cyclic successions of carbonate-evaporite sequences were deposited reaching original thicknesses of up to 2000 m at the time of deposition (Peryt et al., 2010). In total, seven cycles can be distinguished in the North German Zechstein deposits (cf. Peryt et al., 2010; Strohmenger et al., 1998; Tucker, 1991), of which the studied carbonates belong to the second cycle (Stassfurt) (Fig. 4.2).

The formation of those depositional cycles is reflected by sea level fluctuations that can be organized as sequence stratigraphic intervals. Strohmenger et al. (1996) propose eight subdivisions (ZS1 to ZS8) in third-order depositional sequences to classify the German Zechstein sequences. The studied Ca2 carbonates belong to the third sequence (ZS3) of that scheme, representing both highstand and transgressive systems tracts (Peryt et al., 2010; Strohmenger et al., 1996; Strohmenger et al., 1998) (Fig. 4.2).

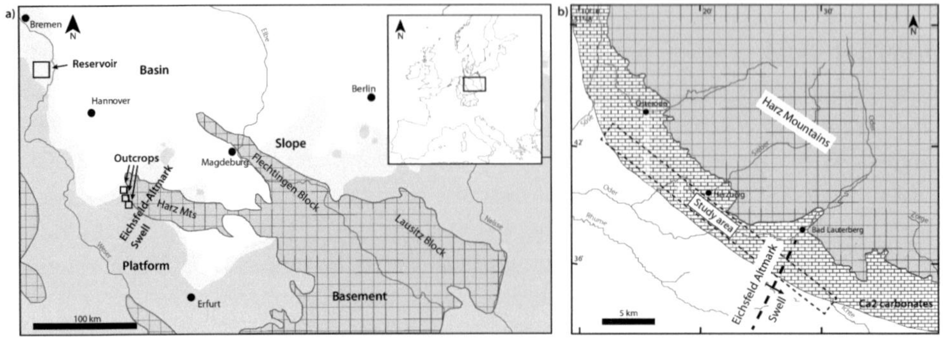

Figure 4.1. a) Distribution map of the basin to platform facies along the southern margin of the Southern Permian Basin including the studied outcrop locations and the location of a gas-producing Ca2 reservoir (modified after Peryt et al., 2010; Pharaoh et al., 2010; Strohmenger et al., 1996). b) Close-up of the studied area, SW of the Harz mountains, with quarry 1 located in the NW part of the marked area, quarry 3 in the SE, and quarry 2 in between, showing the distribution of the Ca2 carbonates and the assumed central position of the Eichsfeld-Altmark Swell (modified after Herrmann, 1956; Paul, 2014).

During the Late Permian, the southern margin of the Southern Permian Basin was characterized by several NNE – SSW striking paleo-highs, at whose margins the Zechstein deposits developed (Pharaoh et al., 2010). The studied carbonates were deposited at the Eichsfeld-Altmark Swell (Herrmann, 1956; Paul, 1993) (Fig. 4.1b). Thus, different carbonate facies depending on their deposition on the platform, slope, or in the basin can be distinguished (Fig. 4.2b). Strohmenger and Strauss (1996) and Strohmenger et al. (1998) give detailed descriptions of those depositional types and associated facies groups. In general, platform facies include sub-, inter-, and supratidal milieus with depositional settings from ooid bars and shoals (subtidal), tidal flats (intertidal), and sabkhas (supratidal) (Peryt et al., 2010; Strohmenger et al., 1998; Strohmenger and Strauss, 1996). Thus, mainly packstones and grainstones and minor laminated wackestones occur in the platform facies (Peryt et al., 2010; Reijers, 2012). Slope facies vary with regard to their position on the slope (cf. Clark, 1986). While proximal slope facies contain turbidite or slump deposits that infiltrate those mainly massive to wavy-bedded mud- to wackestones (Paul, 1991; Peryt et al., 2010; Reijers, 2012; Strohmenger et al., 1998; Strohmenger and Strauss, 1996), more distal slope facies represent laminated to centimeter-bedded mudstones (Reijers, 2012; Strohmenger and Strauss, 1996). Basinal carbonate facies are generally finely laminated carbonates (Strohmenger et al., 1998; Strohmenger and Strauss, 1996), that are rich in organic contents and thus, have a source rock potential (Peryt et al., 2010).

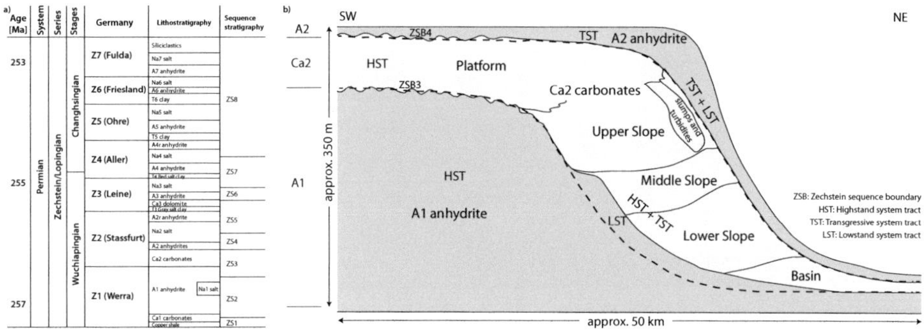

Figure 4.2. a) Stratigraphy of the German Zechstein with lithostratigraphic and sequence stratigraphic classifications (modified after Peryt et al., 2010; Strohmenger et al., 1996). b) Schematic profile of the depositional facies of the Ca2 carbonates on the former margin of the Southern Permian Basin including sequence stratigraphic classes as exposed in the studied area of the southern Harz mountains (modified after Strohmenger et al., 1996).

4.4 Materials and methods

In the three studied quarries, lithological descriptions of the vertical quarry stratigraphy were supplemented by spectral gamma ray values measured with a hand-held, auto-stabilizing gamma ray spectrometer RS-230 BGO Super-SPEC with a 103 cm^3 bismuth germanate (BGO) detector. Measuring positions along vertical profiles in the quarries were set at 30 cm steps, and the integration time at each position was three minutes.

Rock sampling in the outcrops was carried out by drilling plugs parallel to bedding by using a water-cooled core drill. All plugs are 2.54 cm (1 inch) in diameter and sample preparation included oven-drying at 60°C for a minimum of one week.

Petrophysical measurements were performed on 75 (quarry 1 n=32, quarry 2 n=6, and quarry 3 n=37) carbonate samples, and supplemented by data of 15 subsurface samples. For helium pycnometry, the semi-automated AccuPyc II 1340 by micromeritics® was used to determine the porosity of the plug samples. Permeability measurements were done using an isostatic flow cell with single-phase gas flow. Helium (η=1.98*10-5 Pa*s) is used as permeant and confining pressures are created by water. In accordance to plug sample data from the reference well, confining pressures of 12 MPa were applied onto the samples. The permeability coefficient k is determined using Darcy's law (e.g., Amann-Hildenbrand et al., 2016), and given in the unit 'milli-Darcy' (mD) in this study. A total of 51 thin sections (quarry 1 n=27, quarry 2 n=4, and quarry 3 n=20) were prepared from the associated plug caps. The 30 µm polished thin sections were prepared with blue-dyed epoxy resin highlighting open pore space, and samples of quarry 1 and 3 were stained with Alizarin Red S and potassium ferricyanide to distinguish the present carbonate phases.

Petrographic analysis was performed with a Leica DMLP microscope fitted with a Jenoptik Progres Gryphax Subra digital camera.

4.5 Results

4.5.1 Ca2 carbonates of Quarry 1

In the first studied quarry (cf. Becker et al., 2018), a vertical succession of Ca2 carbonates of more than 25 m is exposed (Fig. 4.3). The carbonates are of pale gray color and mainly dolomitic composition. Regarding the lithology of the 25 m high vertical profile, the carbonates can be described from bottom to top as follows:

The lowermost 20 m of carbonates contain thinly-bedded mud- to wackestones (see Dunham classification based on depositional texture, e.g., Flügel, 2010). Bed thicknesses increase upwards from around 3-10 cm in the lowermost 5 m of this unit, that are interlayered with clay-rich beds of up to 5 mm thickness, to a thickness of around 15-30 cm in the upper parts reaching up to 40 cm towards the top of this unit. This coarsening-upwards trend is accompanied by an increase in the original fossil content of those carbonates. While the bottommost 5 to 7 m are described as mudstones, the upper part of the thinly-bedded carbonates features wackestones, which is confirmed by observations of thin sections (Fig. 4.3). A fine-grained dolomite matrix with no visible open pore space is present in the mudstone interval (no. 3, Fig. 4.3). In wackestone samples, remnants of shell fragments (no. 2, Fig. 4.3), ostracods, or foraminifera occur within the fine-grained dolomite matrix. The fossils are replaced by dolomite crystals, and sometimes contain open porosity.

On top of that 20 m thick, thinly-bedded carbonate unit, 5 to 7 m of massive dolomites are exposed. Locally, this unit appears channelized (Fig. 4.3). Thin sections reveal a crystalline dolomite matrix, in which shadows of former ooids and few open pores can be observed (no. 1, Fig. 4.3). Thus, those massive dolomites represent pack- to grainstones.

The top of the exposed carbonates contains a limestone unit of around 1 m thickness. It is described as a lime-mudstone. No fossils were observed (Fig. 4.3).

The Ca2 carbonates are overlain by more than 5 m of A2 anhydrites in the NW part of the quarry, and their red clay A2 equivalent in the SE (Fig. 4.3).

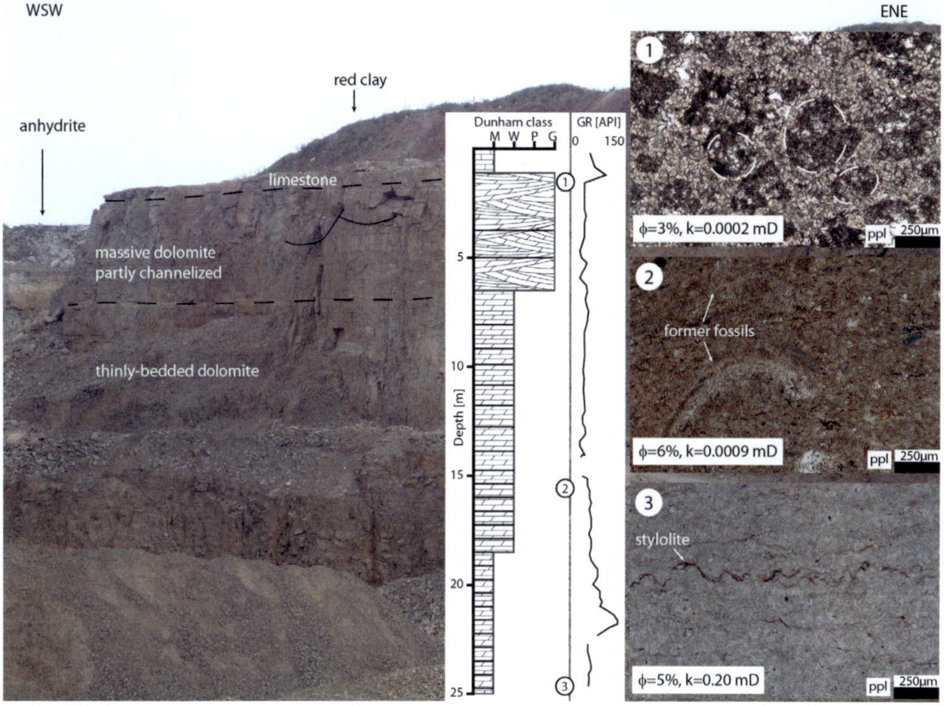

Figure 4.3. Overview of the Ca2 stratigraphy exposed in quarry 1. Upwards-increasing bed thickness is schematically sketched in the lithological column showing the upwards transition from mud-, and wacke-, to pack- and grainstones in the dolomites overlain by a lime-mudstone unit. Gamma ray signals (GR, in API) show increased values in the mudstone sections. Thin section micrographs of samples taken from the different units and their relative stratigraphic height are shown. 1: Shadows of former ooids in crystalline dolomite representing grainstones of the massive dolomite unit (plane polarized light, Ue_2.3). 2: Remnants of shell fragments replaced by dolomite cements in a fine-grained dolomite matrix representing wackestones of the thinly-bedded unit (plane polarized light, Ue_03). 3: Bedding-parallel stylolite in a fine-grained dolomite matrix representing mudstones of the lowermost thinly-bedded unit (plane polarized light, Ue_2.21). Porosity (ϕ) and permeability (k, at 12 MPa confining pressure) values are given for each thin-section sample.

4.5.2 Ca2 carbonates of Quarry 2

The second quarry of this study vertically exposes approximately 41 m of Ca2 carbonates (Fig. 4.4). The main part of the carbonates is of dolomitic composition. From bottom to top, the lithology can be described as follows:

The lowermost 3 m, which are only exposed in the NW part of the quarry, contain thinly-bedded dolo-wackestones with bed thicknesses of around 20 cm. They show pale gray colors with brownish weathering colors.

Thin section micrographs show fine-grained, partly crystalline dolomite matrix with few fossil remnants but no visible open pore space (no. 3, Fig. 4.4).

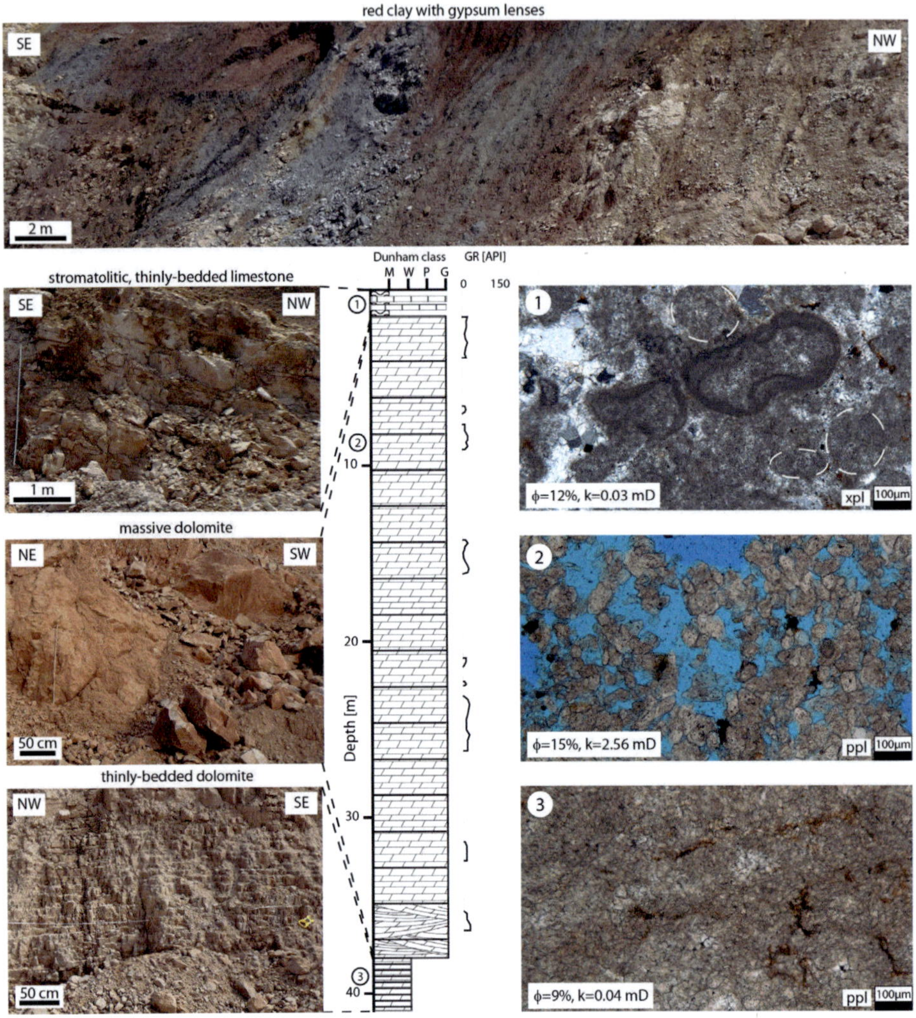

Figure 4.4. Overview of the Ca2 stratigraphy exposed in quarry 2. With an exposed thickness of 35 m, dolo-grainstones represent the main unit, that show generally low gamma ray signals (GR, in API). Thin section micrographs of samples taken from the different units and their relative stratigraphic height are shown. 1: Shadows of former ooids within carbonate crystals representing the top limestone unit (cross-polarized light, SF_03). 2: Dolomite crystals with a lot of open pore space (blue) (plane polarized light, SF_04). 3: Partly crystalline. fine-gained dolomite matrix (plane polarized light, SF_01). Porosity (ϕ) and permeability (k, at 12 MPa confining pressure) values are given for each thin-section sample.

The overlying up to more than 35 m thick dolo-grainstones are gray but show brown weathering colors. In the lowermost 6 m, the carbonates appear partly channelized, but the main part of this unit occurs as massive and structureless. In contrast to the underlying mud- to wackestones, a high amount of secondary open pores is already macroscopically visible. Thin section analyses confirm this observation indicating lots of open pore space between the dolomite crystals (no. 2, Fig. 4.4).

The massive dolomites are overlain by an up to 1.5 m thick limestone unit. It features stromatolitic layers and thin sections show the presence of former ooids in a recrystallized carbonate matrix (no. 1, Fig. 4.4).

The A2 red clay equivalent overlies the Ca2 carbonates with thicknesses of more than 5 m. Locally, gypsum lenses occur filling former karst structures within the red clay (Fig. 4.4).

4.5.3 Ca2 carbonates of Quarry 3

In the third quarry, a vertical succession of more than 17 m of Ca2 carbonates is exposed (Fig. 4.5). As in the two other quarries, the main part of the studied carbonates is of dolomitic composition. Regarding a stacked vertical profile, the lithology can be described from bottom to top as follows:

The lowermost 5.5 m represent decimeter-scale-bedded carbonates (Fig. 4.5). The bottom parts of this unit feature dolo-wackestones that are only locally exposed with thicknesses of around 2.5 m. They are thinly bedded with bed thicknesses of up to 30 cm. The following 3 m show an increase of bed thickness up to around 1 m in dolo-grainstones. The grainstones grade into packstones, that show bed thicknesses of up to 30 cm in the overlying 2 m. Macroscopically, secondary pore space can be observed, where ooids are dissolved in the grainstones.

On top of the bedded carbonates, up to 9 m of channelized dolo-grain- to packstones are exposed. Thin sections reveal the presence of former ooids within a crystalline dolomite matrix (no. 1 and 3, Fig. 4.5). Calcite cements can be present in between the dolomite crystals (no. 3, Fig. 4.5) or partly replacing an entire fossil (no. 2, Fig. 4.5).

The channelized dolo-packstones are locally interlayered with an approximately 1 m thick unit of stromatolitic limestones, that again occur on top of the exposed dolomites with an exposed thickness of approximately 1 m (Fig. 4.5).

In this quarry, the exposed Ca2 carbonates are overlain with the A2 red clay equivalent and neither anhydrite nor gypsum are present.

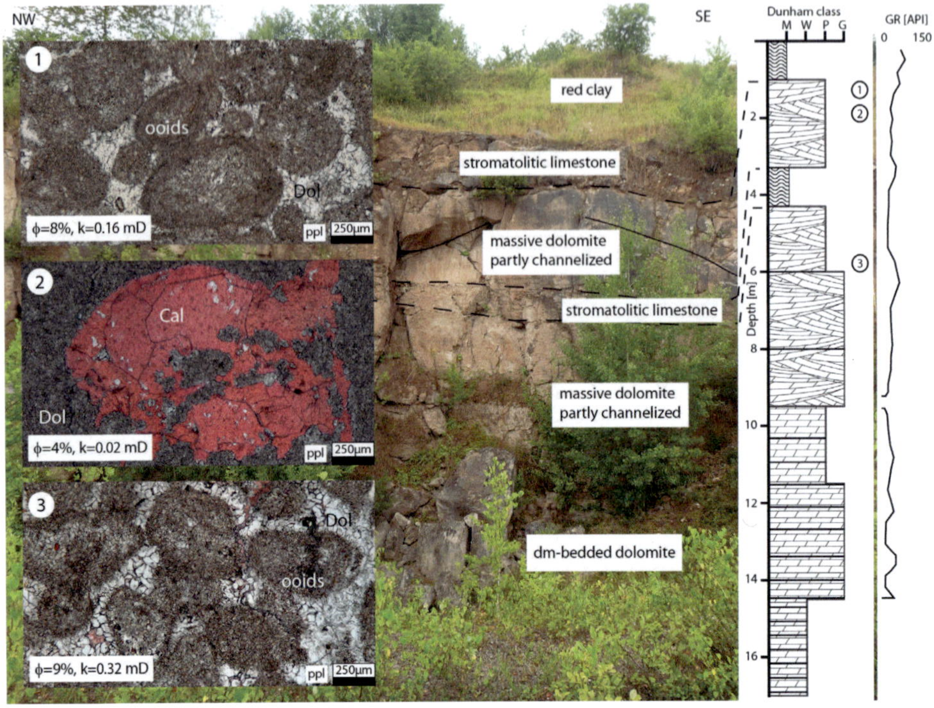

Figure 4.5. Overview of the Ca2 stratigraphy exposed in quarry 3. dm-bedded dolomites are overlain by channelized dolomites and stromatolitic limestones in the top parts of the exposed carbonates. Gamma ray signals slightly increase towards the top unit. Thin section micrographs of samples taken from the main channelized units and their relative stratigraphic height are shown. 1: Remnants of former ooids within dolomite crystals (cross-polarized light, NP_1). 2: Calcite cement replacing a former fossil (plane polarized light, NP_2a). 3: Ooid remnants within dolomite crystals with partly calcite cements in between the crystals (plane polarized light, NP_11). Porosity (ϕ) and permeability (k, at 12 MPa confining pressure) values are given for each thin-section sample.

4.5.4 Petrophysical properties

The permeability characteristics of plug samples drilled in all three quarries generally show a wide range of permeability values in the order of 0.00001 to 10 mD and porosity values ranging from 1 to 21% (Fig. 4.6). With regard to the single quarries, samples of quarry 1 show the lowest porosity and permeability characteristics, while samples of quarry 2 show highest values. For quarry 1, an average permeability of 0.03 mD and porosity of 6% are measured. The average permeability values of the carbonates of quarry 3 are increased by one order of magnitude to 0.7 mD and likewise, the average porosity is higher at approximately 9%. The samples of quarry 2 show a further increase in permeability by one order of magnitude resulting in an average permeability values of 8.87 mD. In similitude, porosities are increased to an average value of approximately 16%. Compared to samples of a reference well of a gas-producing Ca2 field 130 km towards the NW (cf. Field X in Koehrer et al., 2016; see Fig. 1), permeability values of the reservoir carbonates are in a similar range, showing an average of 2.29 mD. Porosities are in a slightly lower range than outcrop samples from 1 to 7% with an average value of approximately 4% similar to quarry 1.

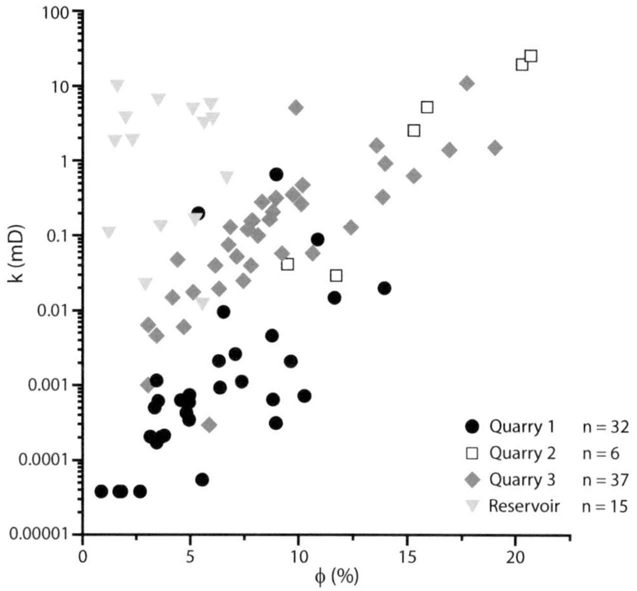

Figure 4.6. Permeability-porosity cross-plot of 75 outcrop and 15 subsurface samples. Subsurface samples show higher permeabilities than outcrop samples, but porosities in a similar range.

4.6 Discussion

4.6.1 Gross depositional environment interpretation

Observations on the Ca2 carbonates of the three quarries reveal differences and similarities of the exposed rocks. The lowermost unit in each quarry contains cm- to dm- bedded dolomitic carbonates, that are characterized as mud- or wackestones (see Figs. 4.3, 4.4, and 4.5). Measured gamma ray values appear increased in this unit (see Fig. 4.3). Such characteristics can be linked to a depositional environment on a proximal part of the slope at water depths too deep for the formation of fossils (cf. Clark, 1986; Paul, 1991; Reijers, 2012; Strohmenger and Strauss, 1996). Reworked particles or fossils that occur in the observed wackestones could have been re-deposited from the platform by slump or turbidite events (cf. Paul, 1991; Peryt et al., 2010; Reijers, 2012; Strohmenger et al., 1998; Strohmenger and Strauss, 1996).

As these bedded wackestones are overlain by a massive pack- to grainstone unit in all three quarries, a transition of a slope to a platform environment is indicated. The former presence of ooids, as observed in thin sections (see Figs. 4.3, 4.4, and 4.5), together with the occasional occurrence of channelized grainstones hints to a higher energy setting of a subtidal, platform margin shoal setting (Peryt et al., 2010; Reijers, 2012; Strohmenger et al., 1998; Strohmenger and Strauss, 1996). The transition from the underlying slope-related carbonates to the platform is further indicated by lower measured gamma ray values (see Figs. 4.3,4.4, and 4.5).

In quarry 2 and 3, a third unit featuring stromatolitic limestones is present on top of the platform margin related units (see Figs. 4.4 and 4.5). An increase in the measured gamma ray signal additionally indicates a transition to another unit (see Fig. 4.5). Stromatolites hint to a formation within an intertidal, inner platform environment (Reijers, 2012). In the subsurface, this type of platform facies would be expected to be dolomitized (e.g., Reijers, 2012; Strohmenger et al., 1996). Intense calcitization (dedolomitization) can result from the migration of fluids enriched in carbon dioxide and a calcium source (Clark, 1980). They are reported to be provided by a combination of the preceding process of anhydritization and the release of CO_2 from organic matter deposited on the distal slope and in the basinal parts (Clark, 1980; Reijers, 2012). As greater thicknesses of the A1 and A2 anhydrites were deposited towards the basin (e.g., Paul, 1993, 2014; Strohmenger et al., 1996) and the migration pathways of the CO_2 enriched fluids would have been the longest towards the inner platform, this would suggest a lower degree of calcitization related to an inner platform setting. However, for the outcrop carbonates the topmost unit represents the parts of the quarries that have been longest exposed to the surface, and thus, to meteoric waters that fulfill the requirements to replace dolomite by calcite (Clark, 1980).

In general, this transition of slope to platform margin and then inner platform deposits hints to a shallowing upwards and prograding carbonate slope in a highstand systems tract of a Zechstein cycle (Fig. 4.7a) (cf. Strohmenger et al., 1996). Above the following sequence boundary, another relative rise in sea level is indicated by the overlying deposits of the A2 anhydrite and red clay on top of the Ca2 carbonates in the studied quarries (see Figs. 4.3, 4.4, and 4.5). However, such a direct correlation does not occur basin-wide for the Ca2/A2 transition, as locally, this boundary can be gradational and interpretations then state the sequence boundary within the A2 (Strohmenger et al., 1996).

4.6.2 Correlation across the Eichsfeld-Altmark Swell

Zechstein deposition can be generally related to several NNE – SSW striking paleo-highs along the southern margin of the Southern Permian Basin (Pharaoh et al., 2010). In case of the studied Zechstein carbonates, they were deposited along the Eichsfeld-Altmark Swell (Herrmann, 1956; Paul, 1993, 2014). In the previous chapter, it is shown that all carbonates exposed in the three quarries can be subdivided into different gross depositional environment types containing the slope-, platform margin-, and inner platform environment. Using that information and the relative positions of the quarries along the Eichsfeld-Altmark Swell, the studied carbonates can be correlated (Fig. 4.7).

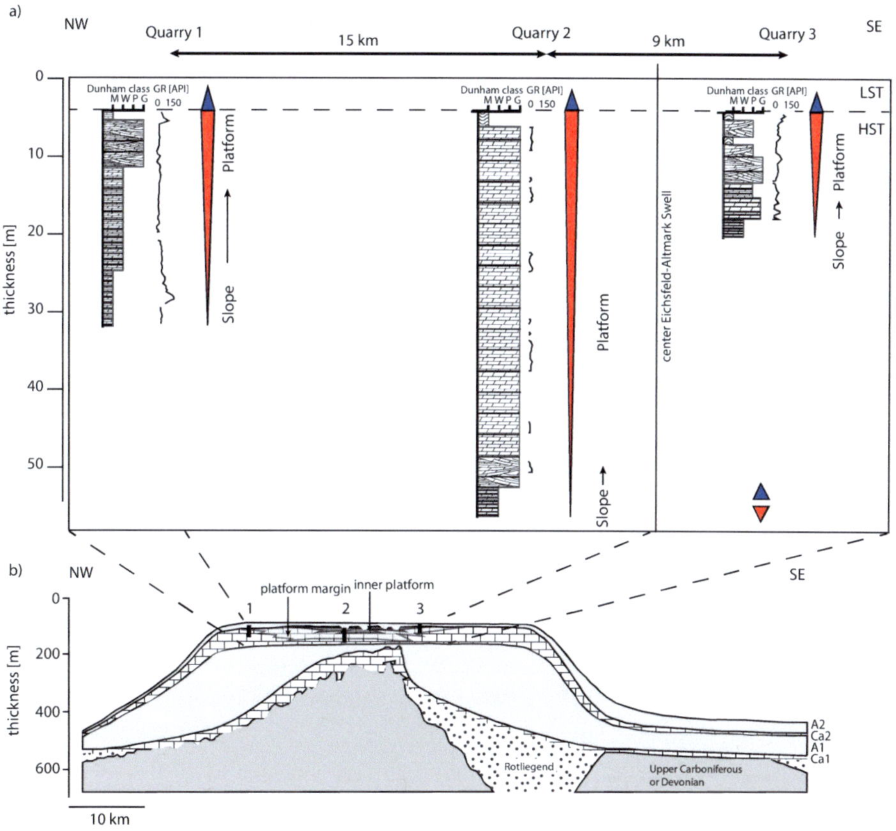

Figure 4.7. a) Correlation of the three studied quarries on the basis of the Ca2 carbonate/A2 anhydrite transition representing the sequence boundary (SB) between a shallowing upwards highstand systems tract (HST) and a deepening upwards lowstand systems tract (LST). b) Cross-section of the Eichsfeld-Altmark Swell (see Fig. 4.1b for location) with the relative positions of the studied quarries (modified after Paul, 1993, 2014).

While slope GDEs are exposed with greatest thicknesses of around 20 m in quarry 1, only around 3 m of this type is present in quarry 2 and 3. In contrast, quarry 2 contains the thickest deposition of platform margin carbonates with more than 35 m. In quarry 1 and 3, those carbonates reach thicknesses of less than 10 m. While carbonates related to the inner platform are absent in quarry 1, they occur both in quarry 2 and 3. This knowledge, together with the assignment of the Ca2/A2 transition as sequence boundary, is used to correlate the quarries in relative distance to each other (Fig. 4.7a). With the relative position of the quarries to each other and to the assumed center of the Eichsfeld-Altmark Swell (see Fig. 4.1b), the exposed carbonates can be related to their position on that paleo-high (Fig. 4.7b).

4.6.3 GDE vs reservoir characteristics

For the Ca2 carbonates, it is widely accepted that porosity characteristics can be related to the facies distribution (e.g., Clark, 1986; Reijers, 2012; Strohmenger et al., 1996; Strohmenger et al., 1998). As the carbonates of the three studied quarries can be distinguished into their types of depositional environment, measured porosity and permeability characteristics can be accordingly related to those analyzed slope, platform margin, and inner platform GDE types (Fig. 4.8).

Results of this study indicate lowest reservoir quality for the slope GDE type with average porosity values of approximately 6% (1-15%) and permeabilities in the range of 0.00001 to 1 mD. This is in accordance with observations by Strohmenger et al. (1996) who state porosities of up to 15% for the dolomitic slope facies, and with Reijers (2012) who reports expected low permeabilities in the slope facies. Apart from the work of Clark (1986), few actual facies-related permeability values have so far been published. Clark (1986) presents data from a Dutch reservoir close to the German border with permeabilities that usually do not exceed 1 mD in reservoirs of the Ca2 slope facies, though local leaching processes might enhance values to up to 100 mD. Reservoir data of this work might fit to such an additionally leached interval of the slope, but only permeabilities are enhanced while porosities remain in a similar range to the outcrop samples (Fig. 4.8). Increased permeabilities could be related to the occurrence of micro-fractures within the measured plug samples (e.g., Lucia, 2007). Further investigations are necessary to explain the discrepancy in permeability between the outcrop and reservoir slope carbonates, which are beyond the scope of this study. The generally low reservoir quality of the slope facies is related to a combination of few interconnected vuggy pores resulting from a rare deposition of fossils or grains that could have been dissolved during early diagenetic processes (e.g., Strohmenger et al., 1996) and intense calcitization processes in these slope carbonates (cf. Biehl et al., 2016; Clark, 1980; Reijers, 2012). However, secondary porosity can be created due to burial related dolomitization (Strohmenger et al., 1996) and leaching processes associated with decarboxylation of organic matter (Clark, 1980, 1986).

Regarding the inner platform GDE type, reservoir quality is enhanced by one order of magnitude regarding the average permeability (slope: 0.04 mD, inner platform: 0.24 mD) compared to the slope GDE type (Fig. 4.8). Porosities show an increase from approximately 6% (slope) to 8 % (inner platform). Thus, reservoir characteristics are still similarly low, as also in this environment, mainly carbonate mud precipitated resulting in tight rocks with few open, connected pores (cf. Reijers, 2012).

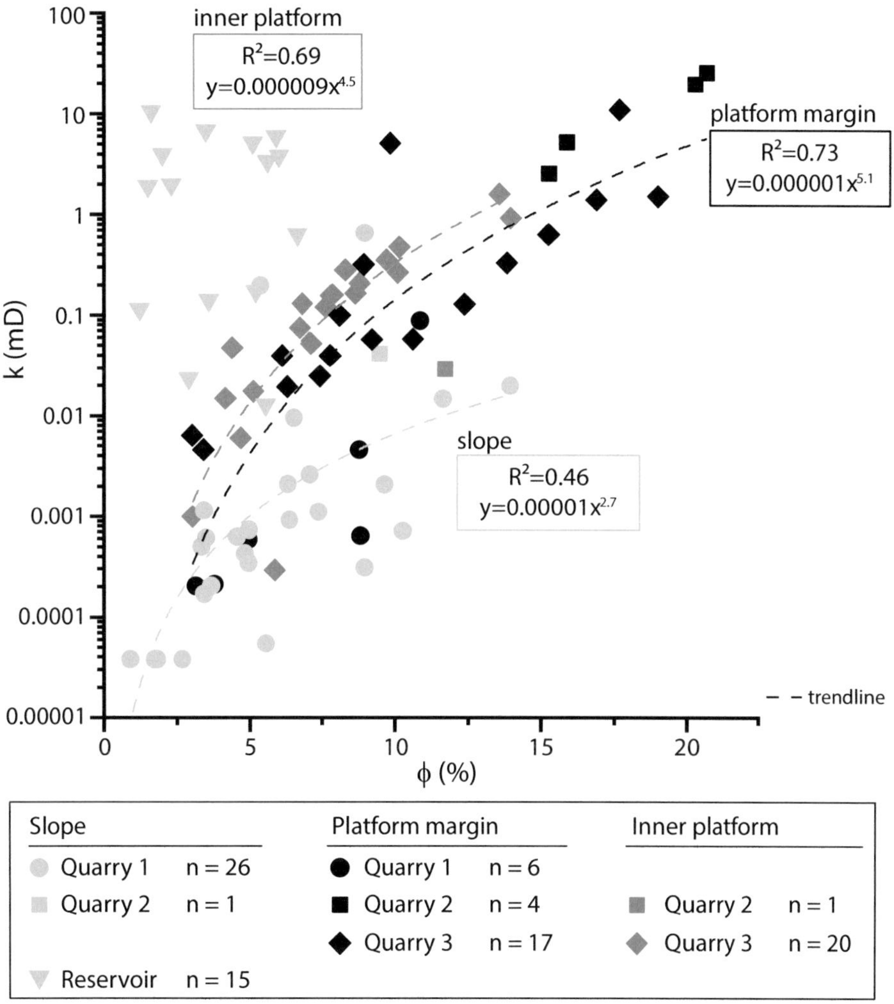

Figure 4.8. Permeability-porosity cross-plot of 75 outcrop and 15 subsurface samples associated to the interpreted GDE type of the sample. Carbonates of the platform margin show best reservoir characteristics compared to the inner platform and slope GDE types. Best correlation coefficients occur for the platform margin GDE type (R^2=0.73), while results of the slope- (R^2=0.46) and inner platform (R^2=0.69) GDE type show wider ranges.

In contrast, carbonates related to the platform margin show the best reservoir quality with average porosity values of approximately 11% (max. 21%) and permeability values up to 25 mD (Fig. 4.8). Strohmenger et al. (1996) state similar results with porosities up to 25% and Clark (1986) shows a similar range of permeabilities with values up to 100 mD. Those good reservoir characteristics are caused by the presence of ooid-rich grainstones forming both primary intergranular and dissolved intragranular open pores (e.g., Clark, 1986; Reijers, 2012). In addition, the low effect of calcitization on those platform carbonates (e.g., Strohmenger et al., 1996) allowed the preservation of intercrystalline porosity resulting from early dolomitization processes (Biehl et al., 2016; Reijers, 2012).

The reservoir characteristics of the studied carbonates correspond to published Ca2 reservoir data (cf. Clark, 1986; Reijers, 2012; Strohmenger et al., 1996) of the southern margin of the Southern Zechstein Basin. Results of this study can therefore be used for reservoir quality modeling (Table 4.1). Porosity-permeability correlations associated to the interpreted GDE type are derived from the outcrop samples (Fig. 8, Table 4.1) and can be incorporated in reservoir modeling. Likewise, gamma ray values appear to be indicative for the type of GDE, which can be used for GDE modeling from well log data (Table 4.1).

Table 4.1. Decision table showing a potential outcrop analog with the corresponding reservoir characteristics (porosity (ϕ), permeability (k), and gamma ray (GR) values) and the derived porosity-permeability correlation for a specific carbonate reservoir GDE type. Gray colors indicate poor analog properties due to limited outcrop exposures.

Reservoir facies		Slope	Platform margin	Inner platform
Potential outcrop analog		Quarry 1	Quarry 2 Quarry 3	Quarry 2 Quarry 3
Average reservoir properties	ϕ (%)	6	11	8
	k (mD)	0.00001-1	0.0001-100	0.001-10
	GR (API)	47	21	38
ϕ /k regression	k=	$0.00001^* \phi^{2.7}$	$0.000001^* \phi^{5.1}$	$0.000009^* \phi^{4.5}$

In general, all studied carbonates represent relatively tight rocks, which indicates the great importance of additional flow pathways as presented by the occurrence of natural open fractures in those reservoirs. Incorporating the studied outcrops as analogs for Ca2 reservoirs and enlarging the field of study to the fracture network provides further crucial information for reservoir modeling in addition to reservoir quality parameters. Careful selection of a correct analog is necessary as fracture mechanics depend on the type of rock, depositional environment, and diagenetic alterations (Gale et al., 2005; Laubach et al.,

2009; Lavenu et al., 2013; Wennberg et al., 2006). Thus, the presented depositional environment distribution across the Eichsfeld-Altmark Swell can assist the selection of an outcrop analog for a distinct reservoir if the expected targeted GDE type is known (Table 4.1). Becker et al. (2018) recently validated this approach by successfully comparing outcrop and reservoir Ca2 carbonates representing the slope GDE type.

4.7 Conclusions

Three studied quarries at the southern border of the Harz mountains, N-Germany expose Zechstein Ca2 carbonates resembling the development of a prograding slope to inner platform environment. They were correlated across the Eichsfeld-Altmark Swell over a lateral distance of 25 km based on the Ca2 carbonate/A2 anhydrite transition representing the sequence boundary (SB) at the top of a shallowing upwards highstand systems tract (HST).

Measured porosity and permeability characteristics are linked to the interpreted gross depositional environment (GDE) indicating best reservoir qualities of the platform margin type (average porosity: 11%, average permeability: 2.74 mD) compared to the slope (average porosity: 6%, average permeability: 0.04 mD) and inner platform (average porosity: 8%, average permeability: 0.24 mD) type. While porosities are similar in subsurface slope samples of a Ca2 reservoir (average 4%), permeabilities appear enhanced in the subsurface samples (average 2.29 mD). Best matrix reservoir qualities are indicated to be related to the center of the studied paleo-high containing the highest proportion of platform carbonate deposits.

Results of the study can be incorporated into conceptual models of Ca2 carbonate deposition and reservoir quality distribution. Porosity-permeability relationships and lateral GDE correlation lengths may be directly used as reservoir modeling input of Ca2 gas fields in the Lower Saxony Basin impacting future exploration and development strategies.

5 Comparing fracture statistics from outcrop and reservoir data using conventional manual and t-LiDAR derived scanlines in Ca2 carbonates from the Southern Permian Basin, Germany

5.1 Abstract

Natural fracture networks strongly control hydrocarbon flow paths in tight carbonate reservoirs. An improved understanding of their geometries regarding orientations and distribution may result in reduced uncertainties in reservoir modeling and well planning. Outcrop analogs provide supplementary information about fracture networks below seismic resolution. We introduce a suitable analog for a gas-producing Zechstein reservoir in the Stassfurt carbonates (Ca2) in the Southern Permian Basin, northern Germany. Dolomite represents the main lithology in outcrop and reservoir rocks, which were deposited on a similar carbonate platform slope environment, and both locations were subject to the same events of diagenetic and tectonic overprint.

The scope of this study is the evaluation of fracture characteristics by a direct comparison of three similar datasets of manual outcrop and digital outcrop data with borehole resistivity image log data from a horizontal gas development well. Manually measured fracture data of the exact same digital scanlines in the outcrop are used to successfully validate terrestrial laser scanning (t-LiDAR). T-LiDAR data is used to generate a high-resolution digital outcrop model, and we introduce a novel workflow, firstly applied to carbonate rocks, to detect fractures from that data set along artificial horizontal wells following E&P industry best practice.

Results of both outcrop and subsurface data suggest W – E orientations for well path planning due to dominant northerly striking open tectonic fracture directions. Higher fracture intensities (P10 values describing the amount of fractures intersecting the scanline) of conventional scanline measurements of 4.3 m^{-1} indicate a bias in the t-LiDAR (P10: 2.6 m^{-1}) dataset inversely related to limitations in fracture detection abilities along the horizontal well. As for the horizontal well (P10: 2.7 m^{-1}), fractures paralleling the borehole appear systematically underrepresented. Fractures with almost perpendicular orientation to the outcrop wall can only be occasionally detected in the t-LiDAR data set due to their limited exposed surface. However, outcrop fracture characteristics are in good correlation with subsurface results and thus, can help reducing uncertainties in reservoir characterization during field development and well planning.

5.2 Introduction

As most carbonate reservoirs are naturally fractured (Lamarche et al., 2012), the understanding of their fracture network geometries is a key to improve reservoir characterization, well planning and thus, production performance (cf. Gillespie et al., 2011; Lamarche et al., 2012; Narr, 1996; Peacock et al., 2003; Wennberg et al., 2006; Wilson et al., 2011; Zeeb et al., 2013). Information from seismic data and potentially few vertical wells provides limited knowledge about the structural inventory of a naturally fractured reservoir during early field development. Many faults and especially fractures are below seismic resolution (Gillespie et al., 2011; Lamarche et al., 2012; Seers and Hodgetts, 2014; Wennberg et al., 2006), and vertical wells tend to underrepresent certain sub-vertical fracture orientations (cf. Narr, 1996; Terzaghi, 1965). Information about the fracture length and aperture is even more difficult to obtain from well data (cf. Ortega et al., 2006; Wennberg et al., 2016; Wilson et al., 2011). Studying outcrop analogs therefore is a means of generating supplementary data to improve the understanding of fracture distributions away from wells and their impact on the fluid flow of the reservoir (cf. Bertotti et al., 2009; Gillespie et al., 2011; Lamarche et al., 2012; Wennberg et al., 2006). While fracture apertures from outcrops only provide a first estimation as actual fracture aperture at depth is controlled by the present-day stress field (Ferrill et al., 1999; Wilson et al., 2011), fracture lengths and heights as well as fracture orientation and spacing adds crucial values for well planning and reservoir modeling.

In recent years, terrestrial LiDAR (light detection and ranging, hereafter referred to as t-LiDAR) techniques have become increasingly important to complement traditional fieldwork studies (e.g., Gigli and Casagli, 2011; Gillespie et al., 2011; Hodgetts, 2013; Seers and Hodgetts, 2014; Wilson et al., 2011; Wüstefeld et al., 2016). In combination with the simultaneous acquisition of digital imagery, additional information such as different rock types can be detected that cannot be resolved in the intensity data obtained by the laser scans (e.g., Hodgetts, 2013). As an advantage to traditional methods, laser scanning allows to collect data from inaccessible areas in a short time resulting in increased sample sizes, and enables both, small and large scale detailed studies of the outcrop back in office, which can be regarded as additional fieldwork time (Gigli and Casagli, 2011; Hodgetts, 2013). Laser scanning can cover a scale of cm to km resolution, and thus, provide data in the order of magnitude of a hydrocarbon field (Hodgetts, 2013; Jones et al., 2009b; Seers and

Hodgetts, 2014). However, the major challenge in using the full potential of this technique is to accurately interpret the dataset and automatically detect discontinuities from the 3D point cloud (e.g., García-Sellés et al., 2011; Kemeny et al., 2006; Laux and Henk, 2015; Riquelme et al., 2014; Seers and Hodgetts, 2014; Wilson et al., 2011). In general, two approaches can be differentiated: manual picking methods or (semi-)automatic segmentation using different algorithms (Gigli and Casagli, 2011; Hodgetts, 2013), based on which several different workflows were introduced (e.g., Cawood et al., 2017; Gaich et al., 2006; García-Sellés et al., 2011; Gigli and Casagli, 2011; Jones et al., 2009a; Kemeny et al., 2006; Laux and Henk, 2015; Riquelme et al., 2014; Seers and Hodgetts, 2014; Sturzenegger and Stead, 2009b; Wilson et al., 2011).

The large number of studies shows that there are different advantages and disadvantages for each workflow, and the question of which to use also depends on the scientific issue of each study. However, the general aim of incorporating outcrop analogs, based on digital or manual data, is to create improved reservoir models. This appears challenging as fracture mechanics are controlled by depositional, diagenetic and tectonic components of each setting (cf. Gale et al., 2005; Lamarche et al., 2012; Laubach et al., 2009; Lavenu et al., 2013; Wennberg et al., 2006). Outcrops are necessarily situated at a lateral distance to their reservoir and have undergone an (additional) uplift to the surface. Thus, the comparability of outcrop and reservoir characteristics depending on the purpose of each study has to be ensured.

We compare an outcrop analog of naturally-fractured Upper Permian Zechstein (Ca2) carbonates with a producing subsurface gas reservoir (cf. Field X in Koehrer et al., 2016) at the southern margin of the Southern Permian Basin onshore North-Germany. We evaluate the quality of two scanlines, each measured by manual and digital means in fractured dolomites. Results are compared with resistivity image log fracture data of a dolomite section of a recently completed horizontal development well in Field X. Therefore, we introduce a digital workflow based on the t-LiDAR data designed to represent an artificial well and (semi-)automatically detect fractures along the well firstly applied to carbonate rocks. This provides the advantages of quicker processing time compared to fracture detection on an entire outcrop wall, and further analysis regarding e.g., the fracture intensities of different sets can be directly determined on that data. To ensure the comparability of outcrop and reservoir fractures, we evaluate their rock composition, depositional environment, and diagenetic history. Furthermore, we expand the comparison of fracture statistics by adding a digitally measured scanline in a limestone wall section from the quarry, that is inaccessible for manual measurements, and compare it to a scanline of a limestone section of the well to show the potential of our introduced workflow for further studies.

5.3 Geological setting

The studied Ührde quarry is located at the southwestern border of the Harz mountains, southwest of the city of Osterode, North Germany (Fig. 5.1), where Upper Permian (Zechstein) Ca2 carbonates are exposed. The area belongs to the southern margin of the former Southern Permian Basin (cf. Pharaoh et al., 2010; Słowakiewicz et al., 2013; Strohmenger et al., 1998), which formed due to crustal extension since the Upper Carboniferous (Scheck-Wenderoth et al., 2008), and showed signs of ongoing thermal subsidence throughout Permian times (Scheck-Wenderoth et al., 2008; Van Wees et al., 2000). Deposition of the studied carbonates took place at the western margin of the 'Eichsfeld-Altmark Swell' (Herrmann, 1956; Paul, 1993), representing one of several NNE – SSW striking paleo-highs influencing the development of early Zechstein deposits along the southern border of the Southern Permian Basin (Pharaoh et al., 2010). This outcrop can be used as a depositional analog for fractured Zechstein carbonate reservoirs approximately 130 km further to the NW in the area of the Lower Saxony Basin (Fig. 5.1), which developed as part of the Southern Permian Basin since the Late Permian (Betz et al., 1987; Binot et al., 1993).

Zechstein deposition encompasses a total of six (cf. Taylor, 1998) to seven (cf. Peryt et al., 2010; Strohmenger et al., 1998; Tucker, 1991) cycles (Z1-Z7, Fig. 5.1). Deposition of the copper shale marks the basin-wide transgressive base of the Zechstein (Peryt et al., 2010; Pharaoh et al., 2010). Subsequently, carbonate-evaporite depositional sequences were deposited. However, only the carbonates of the second cycle (Ca2, Stassfurt carbonate) feature economically important gas reservoirs in the North German Basin with exploration and production activities since 1960 (Karnin et al., 1992; Strohmenger et al., 1996; Strohmenger and Strauss, 1996). Thus, our study focuses on the Ca2 carbonates, representing the transgressive to highstand system tracts of a third-order depositional sequence defined by Strohmenger et al. (1996). The Ca2 carbonates overlie the anhydrites of the first Zechstein cycle (A1, Werra anhydrite), and are overlain by the anhydrites of the second cycle (A2, Basal anhydrite) acting as bottom and top seals for the hydrocarbon accumulations within the Ca2 (Karnin et al., 1996; Peryt et al., 2010; Sannemann et al., 1978; Strohmenger et al., 1996; Strohmenger and Strauss, 1996). Either underlying Carboniferous strata including coal seams or organic-rich basinal carbonates of the Ca2 itself are the source of these reservoirs (Karnin et al., 1996; Strohmenger et al., 1996).

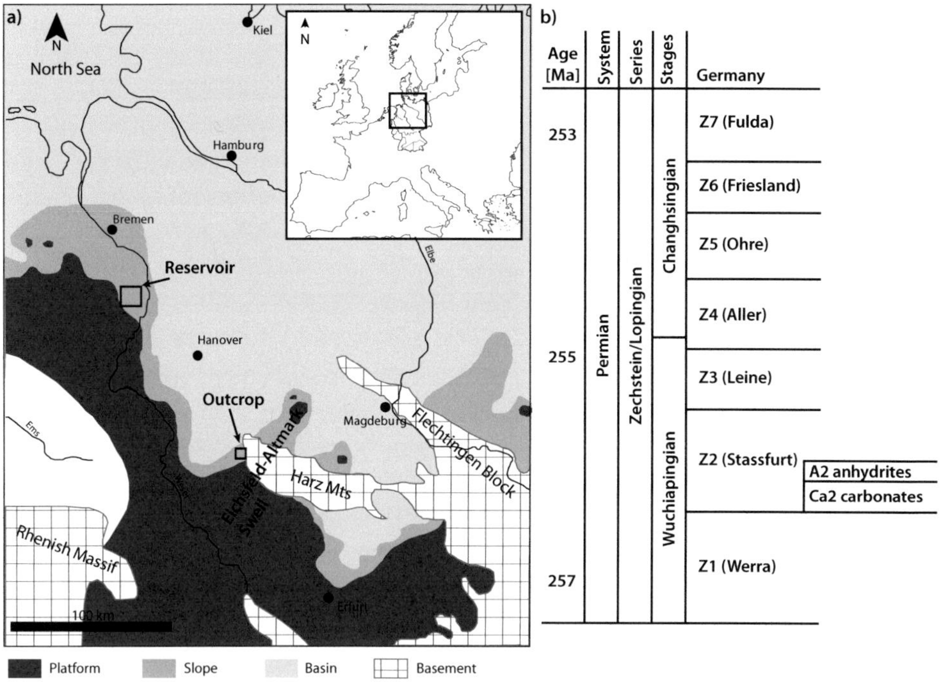

Figure 5.1. a) Location of the studied outcrop and reservoir within the Southern Permian Basin (map modified after Peryt et al., 2010; Pharaoh et al., 2010; Strohmenger et al., 1996). b) Stratigraphy of the Zechstein in Germany with the studied formations of the Ca2 carbonates and A2 anhydrites (modified after Peryt et al., 2010).

Ca2 carbonates developed as deposits on a shallow-marine, rimmed carbonate platform, and thus, main facies can be interpreted as platform, slope, and basin (Fig. 5.1). Several depositional facies types and associations can be distinguished, and are described in detail by Strohmenger and Strauss (1996) and Strohmenger et al. (1998). Depending on the paleogeographic position on the slope, sedimentary features vary (Clark, 1986). Proximal facies are massive to wavy-bedded mud- to wackestones (Peryt et al., 2010; Strohmenger et al., 1998; Strohmenger and Strauss, 1996) featuring slump and turbidite events, which might transport carbonate grains such as ooids or skeletal particles from the platform top to the slope and basin subenvironments (Paul, 1991; Peryt et al., 2010; Reijers, 2012; Strohmenger and Strauss, 1996). More distal slope facies types comprise of dolomitic to calcitic, laminated to centimeter-bedded mudstones which are generally pale in color (Reijers, 2012; Strohmenger and Strauss, 1996). Both, subsurface and outcrop dolomites of this study comprise of slope deposits, deposited during early regression (highstand) of the second carbonate-evaporite cycle.

Structurally, the studied area has been exposed to a NNE – SSW directed extensional post-Variscan rifting regime during the Late Permian and Triassic (Betz et al., 1987; Glennie and Underhill, 1998; Lohr et al., 2007) indicated by W – E to NW – SE striking normal faults and graben systems (Lohr et al., 2007). A rotation of the stress field of NW Europe during the Triassic was caused by crustal separation along with the opening of the Tethys ocean in the Jurassic and resulted in W – E oriented extension (Betz et al., 1987; Franzke et al., 2007; Lohr et al., 2007; Pharaoh et al., 2010; Scheck-Wenderoth et al., 2008). The Upper Jurassic/Lower Cretaceous transition is accompanied by NW-trending transtensional basins and major faults of parallel strike (Betz et al., 1987; Kley and Voigt, 2008; Pharaoh et al., 2010). N – S to NE – SW striking minor faults link the large faults (Kley and Voigt, 2008). In the Cretaceous, W – E to NW – SE compressional directions dominate NW Europe (Betz et al., 1987; Kley and Voigt, 2008; Lohr et al., 2007). As a result of the counter-clockwise rotation during the Africa-Iberia-Europe convergence, a N – S to NE – SW directed compressional regime built-up in northern Germany and resulted in basin inversion in the Upper Cretaceous (Betz et al., 1987; Kley and Voigt, 2008; Lohr et al., 2007; Pharaoh et al., 2010) and the transpressional reactivation of faults (Pharaoh et al., 2010) as strike-slip and thrust faults (Navabpour et al., 2017). Lohr et al. (2007) state an uplift of 1-2 km derived from observed inversion along thrust faults in the Lower Saxony Basin, where the reservoir rocks of this study are located. Uplift around the northern Harz mountains, where the studied outcrop is located, was more than 4 km due to the Northern Harz Boundary Fault (Kockel, 2003; Lohr et al., 2007; Reicherter et al., 2008). Since the late Oligocene, a counterclockwise rotation of NE – SW to NW – SE Alpine compression occurred and is prevailing until today (Betz et al., 1987; Glennie and Underhill, 1998; Kley and Voigt, 2008; Reiter et al., 2016).

5.4 Materials and methods

In the Ührde quarry, rock sampling was carried out using a water-cooled core drill bit and 2.54 cm (1 inch) plugs were drilled parallel to bedding along a WNW – ESE oriented horizontal scanline, as well as representing vertical changes in lithology in the Ührde quarry. Thin sections were prepared from the associated plug caps. The 30 µm polished thin sections were stained with blue-dyed epoxy resin highlighting open pore space, as well as with Alizarin Red S and potassium ferricyanide to distinguish the present carbonate phases. Petrographic analysis was performed with a Leica DMLP microscope fitted with a Jenoptik Progres Gryphax Subra digital camera. Rock strength measurements were performed vertical to bedding on four dolomite rocks sampled at different vertical positions at the Institute of Foundation Engineering, Soil Mechanics, Rock Mechanics and Waterways Construction, RWTH Aachen University.

Fracture patterns were mapped using terrestrial laser scanning and manual scanline measurements. In this study, the term fracture is used to address open joints. The t-LiDAR laser scanner Focus3D X 330 by FARO provides an operating range of up to 330 m. Scanner adjustments of 1/4 resolution and 4x quality were chosen resulting in a definition of 6.136 mm/10 m, i.e. when the scanner is at a distance of 10 m to the quarry wall, scanned points are at a distance of 6.136 mm to each other, and at a 20 m distance, the point distance increases to 12.272 mm. This resolution offers high data quality in a short investigation time with a sampling rate of 976.000 pts/sec. The systematic error on the distance is +/- 2mm. Scan material covers a total area of more than 15,000 m² in the Ührde quarry. To avoid occlusion effects, one outcrop wall section was measured from at least three different locations to detect all surfaces (cf. Sturzenegger and Stead, 2009b; Sturzenegger et al., 2007). Manual fracture data includes a total of 846 scanline measurements containing the fracture orientation as dip direction and dip angle, and the associated fracture location along the scanline to derive the apparent spacing between different fractures (Fig. 5.2). The height of the subvertical fractures, as traceable on the exposed walls, together with information about whether the fractures are constrained to bedding or not, is manually measured as the fracture persistence. Bedding-parallel fractures, that can be related to uplift-related stress release (e.g., Engelder, 1985; Lavenu et al., 2013), have not been manually or digitally measured in this study. As one aim of this study is the comparison of manual and digital fracture data, the defined scanlines represent lines along the quarry walls, along which every fracture is measured. Fracture planes oriented parallel to the scanline were measured at the center of those fracture planes.

Two scanlines were chosen for manual and digital measurement, with their orientations depending on quarry wall geometry. Scanline A is oriented WNW – ESE with a length of 17 m, and scanline B is oriented NNE – SSW with a length of 14.5 m (Fig. 5.3). During manual measurements, the start and endpoints of the scanlines were marked by reference balls, which are used for digital scanning. For scanline A, 68 fractures were measured manually and 60 fractures were identified by t-LiDAR. Analysis of scanline B encompass 74 manual and 42 t-LiDAR fracture measurements. In addition, one scanline was measured for the outcropping limestone, with the same orientation and length as scanline A (WNW – ESE orientation, 17 m), to compare fracture statistics and the applicability of the presented methodology for different lithologies (Fig. 5.3).

Fractures were subdivided in three dip angle classes (gentle: 0-45°, moderate: 46-69°, and steep: 70-90°) which are associated to a thrust regime, conjugate normal faults, and extension (mode I) stress regimes, respectively (cf. Nelson, 2001). In addition, the relative timing of fracture sets was evaluated by mapping 2252 abutting relationships on an approximately 5 m x 4.2 m quarry floor area.

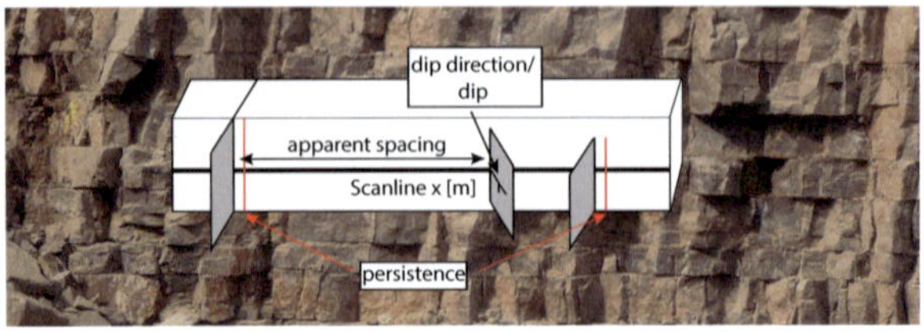

Figure 5.2. The fracture orientation as dip direction and dip angle together with the location of that associated fracture is measured along a scanline to derive the apparent spacing. Persistence is measured as the height of the fracture and relative to the bed thickness.

Subsurface data comprises borehole resistivity image log data (FMI tool from Schlumberger©) of a W – E oriented, 775 m long horizontal well section at a depth of approximately 3.7 km. Aiming for a good comparability with the outcrop analog data of this study, an approximately 30 m long dolomite section of this well was chosen for direct comparison. Fracture data of that section include 81 picked orientations, in total, resistivity image log data provides 1050 open and 98 closed fractures. To compare fracture mechanics with regard to lithology, another 30 m interval of the same well is chosen containing mainly limestone composition. A total of 28 fractures were picked along that interval. Rock-typing, facies analysis, and reservoir quality analysis results are based on four previously drilled vertical wells with 120 m of core material in this field (cf. Field X, Koehrer et al., 2016).

5.4.1 Fracture detection using t-LiDAR scanning

Processing of the terrestrial laser scanning data was performed using the Scene software provided by FARO, leading to a 3D point cloud including both intensity and color images of the quarry (Fig. 5.3). For the next work steps, a point cloud was extracted and the data was edited with the open source CloudCompare software. On the one hand, CloudCompare allows to reduce the amount of data by minimizing, for example, the color codes acquired during scanning. Moreover, segmentation tools are provided, enabling the extraction of scanlines A and B (Fig. 5.3). In the following, a clustering approach (Sibbing et al., 2013), developed with the Center for Computational Engineering Science, RWTH Aachen University, was applied. It rates the neighborhood around each scanned point and combines point data with the same orientation as ellipses, which finally feature the dominant fracture orientation (Wüstefeld, 2018; Wüstefeld et al., 2016) (Fig. 5.3). Wüstefeld (2018) provided a detailed description of the developed fracture detection algorithm. In general, the method of a hierarchical clustering approach successively merges the point data into clusters based on application-specific criterions. The algorithm defines a normal to each point, with the associated eigenvalues mathematically representing an ellipsoid, and a small surface area is generated for a user-defined threshold ratio of those eigenvalues (ratio of smallest eigenvalue (=smallest axis of the ellipsoid) to the square root of the product of the remaining axes). The normal of a cluster initialized at such a point equals the smallest axis of the ellipsoid and neighboring points with equal directions of the normal vectors being merged to that cluster. The merging process of similar clusters stops at a user-defined angle threshold. In a next step, remaining points, as e.g., corner points, are added to the clusters for a user-defined minimal point to plane distance. In addition, the minimum number of points needed for cluster detection is defined (cf. Wüstefeld, 2018). For this study, points and small clusters were merged to one cluster representing the fracture plane according to the following input parameters: The threshold of the ellipsoid axes ratio is set to 0.04, the angle between two neighboring clusters may not exceed 15°, the deviation threshold is set as such that points can have a maximum distance of 1 cm to the cluster, and the minimum number of points needed to generate one cluster, i.e. fracture surface, is set to 15. All values are chosen in a way that correctly displays the fracture surfaces verified by virtual comparison. Wüstefeld (2018) introduced that workflow for synthetic scanlines and in sandstones, that appear less fractured than the studied carbonates. Therefore, the size threshold was reduced for this study to already detect fracture surfaces with a size of $1.2*10^{-3}$ m^2 at an average distance of 20 m between the scanner and the wall (scanner resolution 6.136mm/10m).

The derived fracture data was then loaded in Petrel, which allows to visualize the quarry in 3D and the geometry of fracture data, color-coded according to the dip direction of the fractures (Fig. 5.3). Petrel allows the creation of lateral wells through the fractures, providing virtual scanlines (Fig. 5.3) with the fracture data associated to dip direction, dip angle, and associated apparent spacing. The digitally derived fracture data is comparable to the data measured by manual means. Further processing was done using FracaFlow® of the OpenFlow SuiteTM by BeicipFranlab, which allows to quantitatively describe the fracture data. The generated data was analyzed including the definition of fracture sets and the computation of fracture intensities. For this study, P10 values describing the amount of fractures intersecting the scanline, e.g., per meter (cf. Bisdom et al., 2014; Dershowitz and Herda, 1992), were chosen for easy comparison with resistivity image log interpretations from the well data. If fracture surfaces are not oriented perpendicular to the scanline or well orientation, the derived apparent spacing needs to be corrected, depending on the angle between the fracture set and the scanline/well (Terzaghi, 1965). This correction was applied for the determination of the P10 values for each fracture set based on different strike orientations, evaluated from manual, digital, and well data. Consequently, no such correction can be used to determine a P10 value for all fracture sets combined. In the same way, analyses are done for the manually measured scanline data to check the quality of this method. Furthermore, the subsurface data is likewise digitized and analyzed to compare the results.

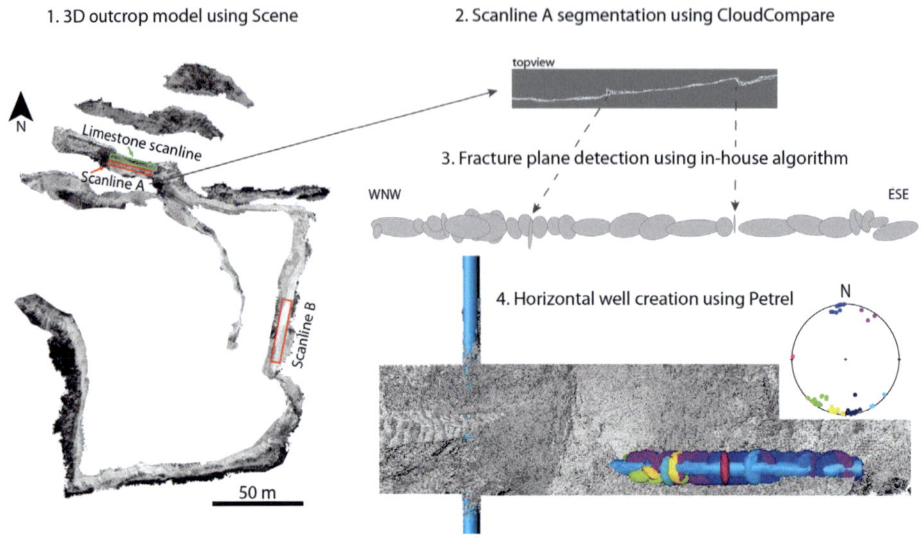

Figure 5.3. Digital workflow from the 3D t-LiDAR derived point cloud outcrop model to an artificial well drilling along the defined scanline A with associated digitally detected fractures. Differently oriented fractures are identified by ellipses and in step 4 color-coded. Stereonet with colored poles displaying the color-code of the fracture planes. Dashed arrows point to ellipses representing that part of the scanline.

5.5 Results

5.5.1 Ca2 carbonates

5.5.1.1 Outcrop

The Ührde quarry vertically exposes more than 25 m of Ca2 carbonates with mainly dolomitic composition (Fig. 5.4a) that show pale gray colors and an overall coarsening-upwards trend. The dolomites have an average E-modulus of 16 GPa and an average uniaxial compressive strength of 161 MPa (Table 5.1).

The lowermost 20 m of exposed carbonates feature thinly-bedded dolomites. A strong smell of petroleum indicates an organic content. In the lowermost 5 m of this package, clay-rich beds of up to 5 mm thickness separate the dolomites, representing mudstones with less than 10% grains (see Dunham classification based on depositional texture, Flügel, 2010) (Fig. 5.4a). The thinly-bedded carbonates show upwards-increasing bed thicknesses. Upwards coarsening is documented by a higher original fossil content visible due to secondary pores inheriting the shape of former bivalves and ostracodes, which is accompanied by a decreasing gamma-ray signal (Fig. 5.4a). Therefore, those carbonates are classified as wackestones.

The following 5 to 7 m of the quarry wall are represented by massive to wavy-bedded carbonates, which appear partly channelized (Fig. 5.4a). The channelized carbonates represent pack- to grainstones of a high-energetic environment.

The topmost 1.1 m of the carbonates consist of lime-mudstones without any visible fossils. A blackish, coarsely crystalline dolomite band of approximately 0.5 m occurs in the top parts, 2 m below the A2 in the NW part of the quarry. This black layer marks the top of the Ca2 in the Osterode area and is referred to as 'anthraconite' (Paul, 2014).

A2 anhydrites overly the carbonates in the NW part of the quarry. In the SE, red clay is present, overlaying the carbonates as A2 equivalent. Both, anhydrite and red clay expose thicknesses of more than 5 m.

Table 5.1. Rock mechanics (uniaxial compressive strength = UCS and E-modulus, given in MPa) of three dolomite rock samples of the Ührde quarry. The relative stratigraphic position of the samples is shown in figure 5.4.

Sample ID	quarry depth [m]	UCS [MPa]	E-modulus [GPa]
Ue 2.12	12.7	233	20
Ue 2.27	23.8	117	13
Ue 2.22	27.1	132	14
Mean		161	16

The measured dolomite scanlines are situated in the upper part of the centimeter-bedded mudstone unit at approximately 16 m quarry depth (Fig. 5.4a, scanline A). The limestone scanline is obtained from the t-LiDAR data set at approximately 0.5 m quarry depth following the orientation and length of scanline A, and thus, represents a wall section inaccessible for manual measurements (Fig. 5.4a, limestone scanline).

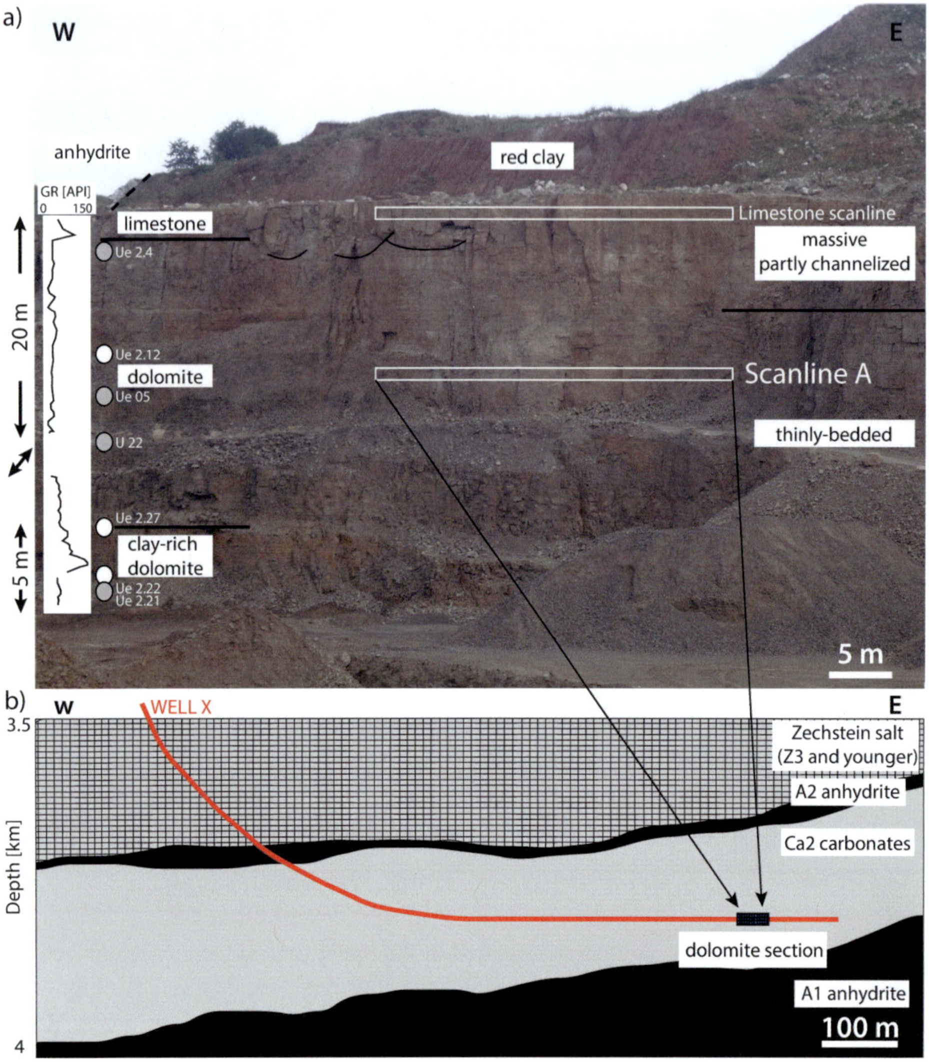

Figure 5.4. a) Overview of the Ührde quarry showing the upwards transition from thinly bedded carbonates to massively bedded carbonates with some channels. Anhydrites (western part of picture) or red clay (eastern part) overly the carbonates. The positions of the measured scanline A and the limestone scanline are marked. The relative stratigraphic position of the samples used to determine the rock strength (white dots, see Table 5.1) and for thin section analyses (gray dots, see Fig. 5.5) is shown. b) Schematic profile of the reservoir setting and well path. The studied dolomite interval of the well is marked, which matches the outcrop scanline A in terms of orientation and length.

Thin section analysis supports the macroscopic observations of the carbonate composition of the Ührde quarry. The staining allows to distinguish dolomitic and calcitic components

(Fig. 5.5b). The matrix is generally composed of dolomite (Fig. 5.5), while dolomite and calcite cementation are present in fossil remnants or filled pores (Fig. 5b and c). The dolomite matrix appears very fine-grained and anhedral including some euhedral dolomite crystals of up to 25 μm size (Fig. 5.5c). When present, fossils appear deformed by compaction and they represent the remnants of foraminifera (Fig. 5.5d), ostracods, and shell fragments. Those fossils provide moldic porosity contributing to the bulk porosity of those carbonates. On the other hand, authigenic cementation processes reducing porosity are visible due to the presence of replacing dolomite and calcite crystals (Fig. 5.5b and c). Quartz is present in the dolomite samples of the massively bedded package (Fig. 5.5e) and shadows of former ooids can be observed in those channelized carbonates (Fig. 5.5f). Intense recrystallization events destroyed the original texture of the carbonates.

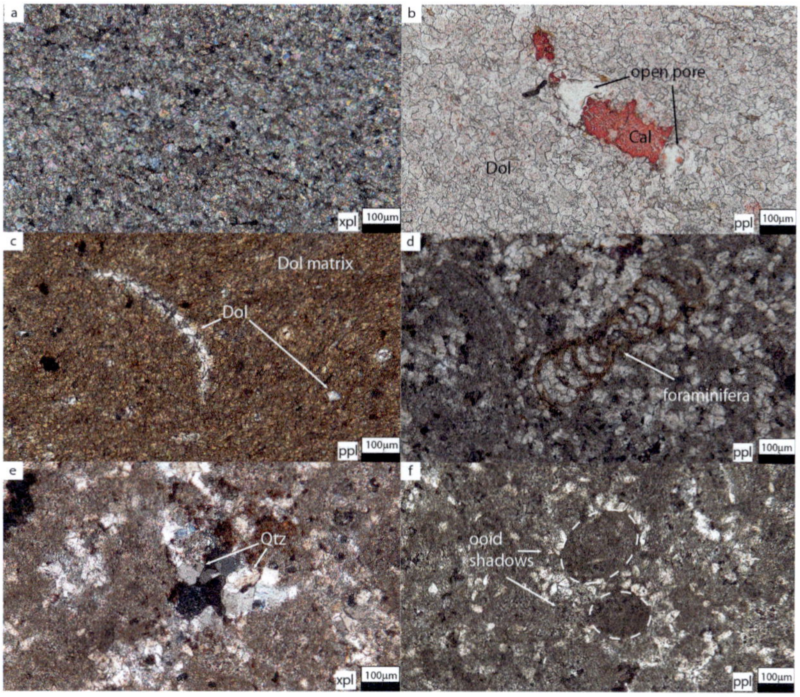

Figure 5.5. Thin section micrographs of dolomite samples from the Ührde quarry. a) Dolomite matrix of a mudstone of the thinly bedded dolomites (cross-polarized light, Ue_05). b) Stained calcite cement filling open pore and dolomite matrix of a mudstone sample (plane polarized light, Ue_2.21). c) Dolomite cement filling potentially former bivalve shell fragment and euhedral dolomite crystals in dolomite matrix of a wackestone (plane polarized light, U_22). d) Remnant of a foraminifera filled with dolomite crystals (plane polarized light, Ue_2.4). e) Quartz occurring in a wackestone of the massively bedded dolomite unit (cross-polarized light, Ue_2.4). f) Shadows of former ooids in a sample of the channelized carbonate unit (plane polarized light, Ue_2.4).

5.5.1.2 Reservoir

Subsurface data comprises of 775 m of drilled Ca2 carbonates. Most of those carbonates represent limestones with dolomitic interlayers. One dolomite interval of approximately 30 m is chosen for the detailed fracture analysis matching the analogous outcrop lithology best (Fig. 5.4b). Another 30 m interval, containing limestone, is chosen to compare the reservoir fractures with regard to their lithology.

Thin sections from core samples of reference wells in Field X (cf. Koehrer et al., 2016) show a fine, anhedral matrix with few sub- to euhedral dolomite crystals. Few remnants of foraminifera and shell fragments are locally replaced by dolomite and calcite cementation.

5.5.2 Fractures

5.5.2.1 Manually measured fracture data

Ca2 dolomites of the Ührde quarry are strongly fractured. Manual compass measurements comprise of 846 open fractures obtained from several dolomite scanlines and their orientations result in the definition of three main fracture sets striking WNW – ESE to W - E, NW – SE, and NNE – SSW to NE – SW based on the maxima in the rose diagram of the fracture data (Fig. 5.6).

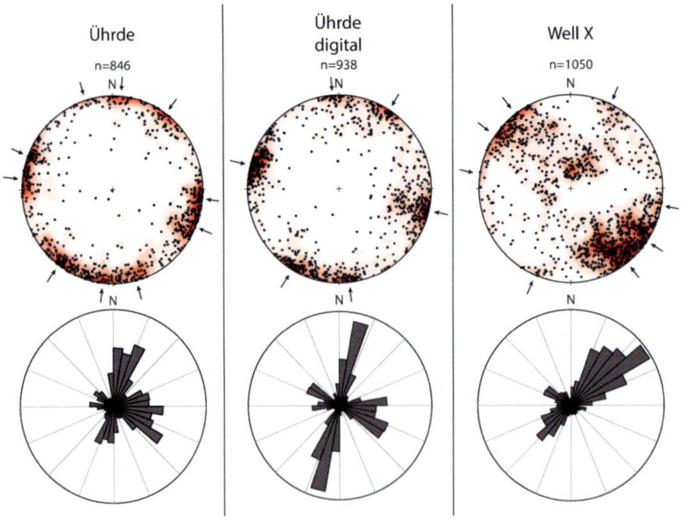

Figure 5.6. Fracture data plotted as poles (lower hemisphere, isolines at 0.5% area, upper row) and rose diagrams (strike, lower row) of the studied outcrop. The first two columns present manual compass measurements and digital t-LiDAR results, and the last column displays reservoir data from resistivity image log analysis. Arrows point to the centers of highest pole densities representing different fracture sets.

To simplify the comparison of the different data sets, a further subdivision into minor sets of e.g., WSW – ENE and WNW – ESE strike directions, that can be derived from the maxima of point density in the stereographic projection (Fig. 5.6), is omitted.

For aperture and vertical persistence (fracture height) statistics, the WSW – ENE and WNW – ESE fracture set are combined as W – E, and NNE – SSW and NE – SW striking sets are referred to as NE – SW striking sets. Measurements on 310 open fractures indicate mean apertures of 1.8 mm with almost no variations among different fracture sets (NE – SW: 1.7 mm, NW – SE: 1.8 mm, W – E: 1.8 mm). Apertures between 0 and 1 mm are most frequent, and opening width of 2-5 mm occur more often than 1-2 mm (Fig. 5.7).

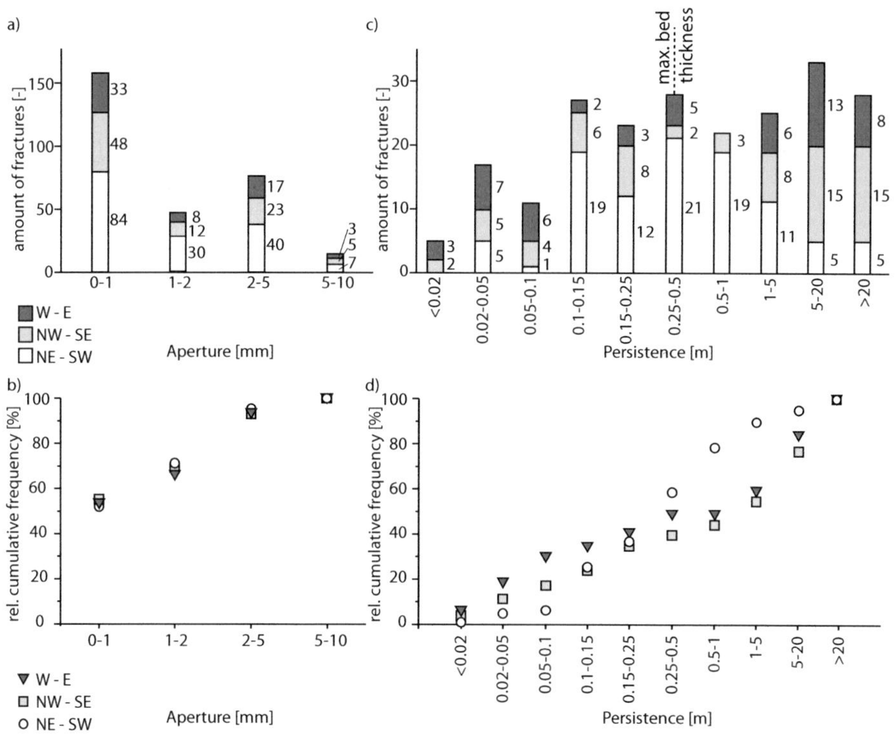

Figure 5.7. Aperture, with total amounts of fractures (a) and their relative cumulative frequencies (b), and persistence, with total amounts of fractures (c) and their cumulative frequencies (d), divided into classes and associated to the three main fracture sets.

Observations on 219 fractures suggest a mean vertical persistence of 4.9 m. In contrast to the aperture, differences occur for the fracture sets with highest values for the NW – SE striking set with average 7.5 m, average 6.4 m for the W – E striking set, and lowest average values of 2.2 m for the NE – SW striking set. Altogether, fractures with a vertical persistence of 5 to 20 m occur most frequently (15%) (Fig. 5.7). Reservoir-relevant fractures exceed bed thickness, which is at a maximum of 30 cm in the bedded dolomite unit. Bedding exceeding fractures mainly occur in the NW – SE and W – E striking fracture sets. Regarding scanline A, the dolomite bed shows a thickness of 12 cm. Out of the 68 manually measured fractures along that scanline, only three fractures are constrained to that bed thickness and those three fractures have apertures of <1 mm.

The majority of fractures in the Ührde quarry are steeply dipping (90%, 760 fractures), while almost 9% (73 fractures) of the fractures are moderate and 1% (13 fractures) show gentle dip angles (Fig. 5.8). Various dip angles occur for all defined fracture sets.

Most fractures areopen and only those were measured along the scanlines and discussed in this study. However, some fractures can be sealed or partly sealed and are mainly filled with dolomite cements. In the Ührde quarry, veins can be best observed on the quarry floor but are only rarely present with W – E and NNE – SSW strike directions. They show small apertures (≤ mm-size) and are very limited in length ranging from few centimeters to decimeter extent (Fig. 5.9a).

Another observation in the outcrop features a black, incohesive organic material on some joint surfaces with NW – SE strike directions (e.g., 032/78°) but not on any other fracture set (Fig. 5.9b).

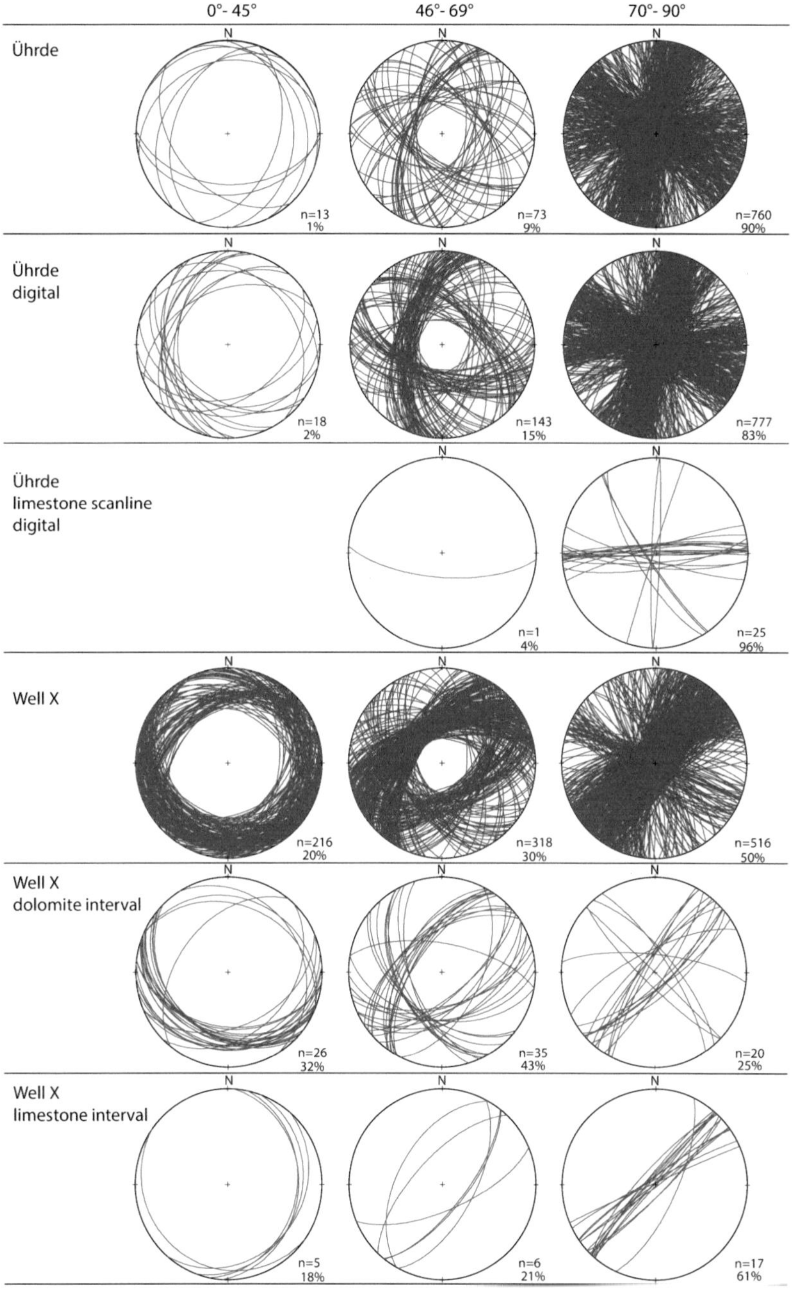

Figure 5.8. Fracture data divided into fracture classes according to the dip angle (gentle: 0-45°, moderate: 46-69°, and steep: 70-90°) of the studied outcrop (complete manual and digital data sets, and digital fracture data of the limestone scanline) and reservoir (complete well, dolomite-, and limestone interval) plotted as great circles in a stereographic projection.

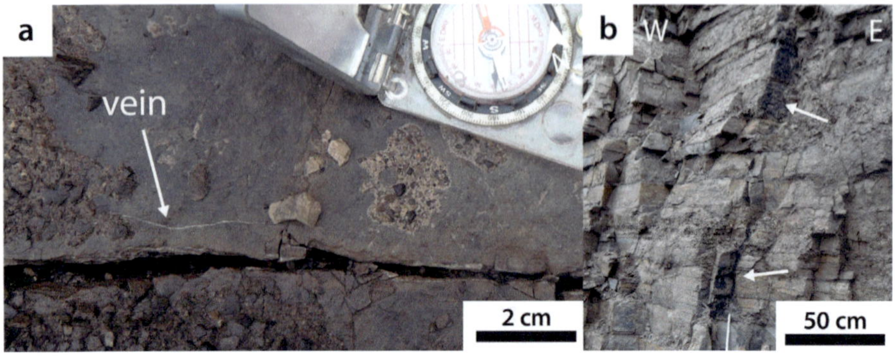

Figure 5.9. a) W – E striking vein on the quarry floor. b) Blackish organic precipitation (white arrows) on few NW – SE striking joint surfaces.

Mapping an approximately 5 m x 4.2 m quarry floor section resulted in the determination of 2252 abutting relationships between differently oriented fractures (Fig. 5.10a). Abutting relationships were plotted in a density plot to get quantitative results (Fig. 5.10b). Fractures with strike directions of 070 to 135° abut fractures with 000 to 030° strike directions with only slightly lower frequencies as fractures with 000 to 030° strike directions are abutting fractures of 070 to 135° strike. Thus, the quantitative results support the observed impression that no unambiguous relative time sequence occurs. However, if observations are restricted to the longest fractures, NNE – SSW to NE – SW fractures might abut at NW – SE striking fractures potentially indicating relative older formation of the latter. A small cluster showing this abutting of younger NE – SW at older NW – SE fractures could support this observation.

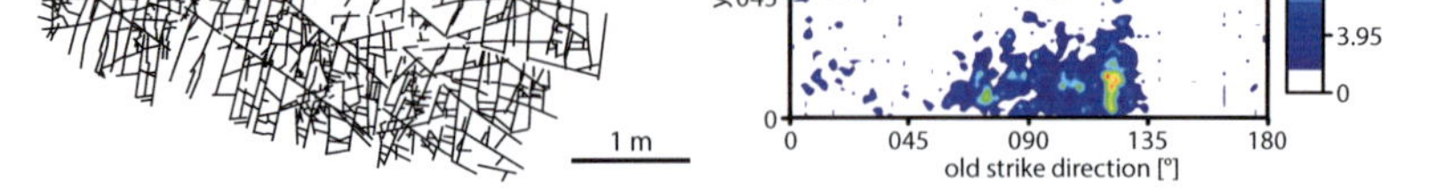

Figure 5.10. a) Sketch showing the fracture patterns of a 5 m x 4.2 m quarry floor section used to derive abutting relationships of differently striking fractures. b) Results are plotted in a Kernel density plot giving the amount of relatively younger fractures abutting against older fractures classified according to their strike directions. A large number of such abutting relationships is indicated by red colors.

5.5.2.2 Digital outcrop fracture data

Using t-LiDAR scanning, a total of 938 fractures were determined along the two different wall orientations in the dolomite unit to observe results in the same order of magnitude as measured manually (846 fractures) and as available from the reservoir data (1050 fractures). Fracture orientations can be divided into three main fracture sets with almost W – E, NW – SE, and NNE – SSW strike directions (Fig. 5.6).

Classifying the fracture data according to their dip angle results in a majority of 777 steeply dipping fractures (83%), 143 oblique fractures (15%), and 18 gently dipping fractures (2%) (Fig. 5.8).

Along the one limestone scanline, 26 fractures were digitally detected which show the same fracture orientations. In contrast to the fractures obtained from the dolomite unit, the dip angles of the limestone fractures are all steep with the exception of one dip angle of 69°, which just falls below the chosen classification limit for steep dip angles.

5.5.2.3 Comparison of manual and digital outcrop fracture data

Regarding the complete data sets (Fig. 5.6 and 5.8), manually and digitally derived fractures show the same orientations (NNE – SSW, NW – SE, and W – E) but with slightly different relative frequencies as visible in the rose diagram and pole isolines (Fig. 5.6). The NNE – SSW striking fracture set appears more dominant in the t-LiDAR data set than in the manual data set. The subdivision in WSW – ENE and WNW – ESE fracture sets of the manual data cannot be detected in the t-LiDAR data set. Regarding the pole densities (Fig. 5.6), a NNE – SSW and NE – SW fracture set can be subdivided in the manual data set, but in the digital data set those fractures show only one center of highest densities and thus, no subdivision in two individual fracture sets. The subdivision into dip angle classes shows similar abundances but more moderately dipping fractures were recorded with the digital (143) than the manual (73) workflow (Fig. 5.8).

In addition, two scanline orientations, one striking WNW – ESE (Scanline A, Fig. 5.3) and one striking N – S (Scanline B, Fig. 5.3), respectively, are chosen to compare individual fracture surfaces derived from the manual and t-LiDAR workflow in the Ührde quarry. Exemplarily, the results of five fracture planes from manual measurements and their digital match are chosen along scanline A (Fig. 5.11). As the same starting point was defined for the manual and digital scanline, the related fractures can be found due to their scanline position. The comparison of the fracture orientations reveals that dip directions have deviations up to 5° and dip angles vary with 2° deviations (Table 5.2).

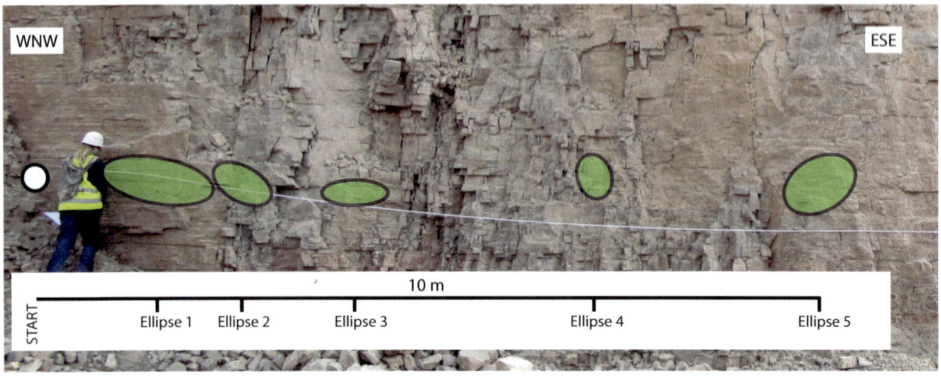

Figure 5.11. Exemplarily digitally derived fracture surfaces are displayed as ellipses along scanline A.

Table 5.2. Comparison of five selected examples shown in Fig. 5.11 showing dip direction and dip angles of manually measured and digitally derived equal fracture surfaces along scanline A.

	manual		digital		deviation	
	dip direction [°]	dip [°]	dip direction [°]	dip [°]	dip direction [°]	dip [°]
Ellipse 1	037	83	033	83	4	0
Ellipse 2	044	80	039	82	5	2
Ellipse 3	346	85	347	86	1	1
Ellipse 4	211	90	216	88	5	2
Ellipse 5	018	86	018	86	0	0

Regarding the complete data set derived from the two scanlines, A and B, 58 fractures were measured manually and detected digitally. Using the defined input parameters for this study, results show deviations between the manually and digitally derived orientations of average 5° for the dip direction and average 4.3° for the dip angle.

However, disregarding the accuracy of the presented method to derive orientations, differences in the detected frequency occur. For scanline A, oriented WNW – ESE, digital results indicate that the NNE – SSW striking fracture set appears underrepresented in contrast to the manually derived results. Similarly, along scanline B, with N – S orientation, the W – E and partly NW – SE fracture sets appear underrepresented compared to the manual scanline data (Fig. 5.12). Thus, fracture sets with parallel or subparallel wall orientation are well detected with the t-LiDAR workflow and those with perpendicular orientation to the wall appear underrepresented. If fracture data from the two wall orientations are combined, results show the same orientations and only small differences in the frequency of each fracture set occur (Fig. 5.12).

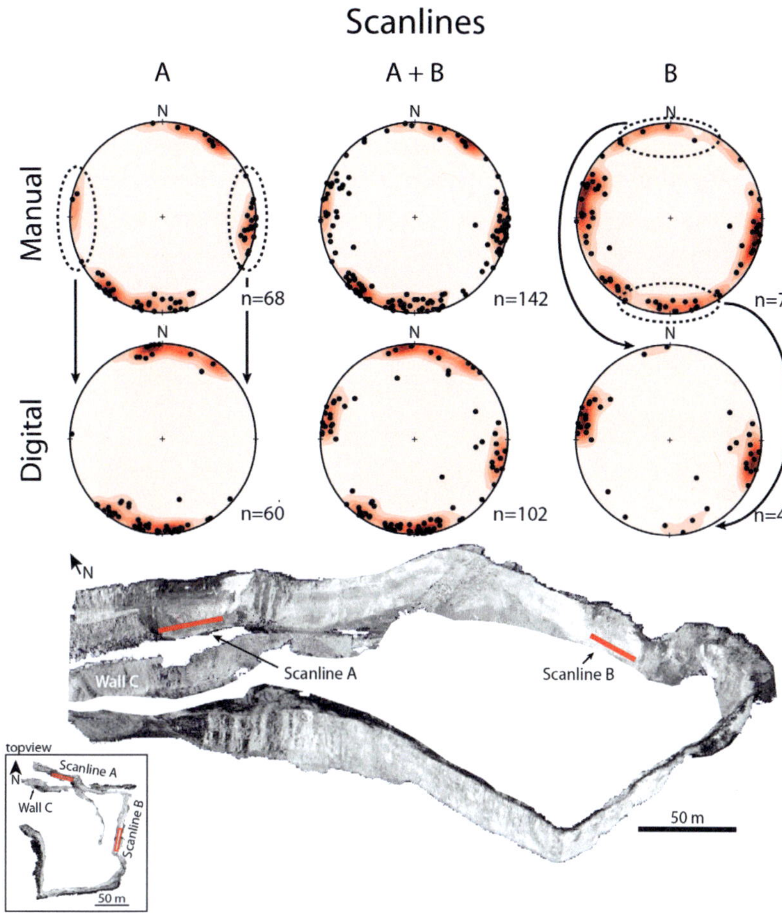

Figure 5.12. Fracture data of the two scanlines (A and B) and the combined set (A+B) plotted as pole points in stereographic projections. Digital fracture data appear underrepresented regarding fracture sets oriented perpendicular to the scanline orientation (see arrows). The position of the two scanlines is shown in the 3D scan of the Ührde quarry.

5.5.2.4 Reservoir fracture data

A total of 1050 open fractures were derived from borehole resistivity image log data from the horizontal well. Two main fracture sets with NE – SW to NNE – SSW and NW – SE strike directions are defined from the rose diagram (Fig. 5.6). The NE – SW set is the best expressed. The NE – SW orientations could be further subdivided into three different sub sets due to the density maxima in the stereographic projection (Fig. 5.6), but this is omitted for the discussion to simplify the comparison with the other data sets.

A division of the fracture orientations with regard to their dip angle (gentle: 0-45°, moderate: 46-69°, and steep: 70-90°) indicates that about 50% (516 fractures) of the subsurface fractures are steeply dipping. In the limestone interval, 61% (17 fractures) of the fractures are steeply dipping. In contrast, evaluating the dolomite section, only approximately 25% (20 fractures) of the fractures are steeply dipping. Here, moderately dipping fractures occur most frequently with 43% (35 fractures) and also gently dipping fractures are very common with 32% (26 fractures). Regarding the entire well, the moderately dipping fractures still occur with 30% (318 fractures), and the gently dipping fractures account for about 20% (216 fractures) thereto. In the limestone interval of the well, 21% (6 fractures) of the fractures are moderately dipping and 18% (5 fractures) show gentle dip angles (Fig. 5.8).

5.5.2.5 Fracture density variations

In a next step, linear fracture intensities (P10) of outcrop and subsurface data are compared (cf. Bisdom et al., 2014; Dershowitz and Herda, 1992). In order to compare the manually and digitally derived fracture data, the most frequent sets and the ones apparently underrepresented in the digital data are used. For scanline A, those are the NW – SE (most frequent) and NNE – SSW (underrepresented in t-LiDAR data) striking sets and for scanline B, the NNE – SSW (most frequent) and W – E (underrepresented in t-LiDAR data) striking sets (Fig. 5.13).

For the NW – SE fracture set of scanline A, manually measured fractures result in a P10 value of 1.8 m^{-1}. Regarding the digitally derived scanline, it shows the same P10 value of 1.8 m^{-1}. For the NNE – SSW oriented fracture set, P10 values show differences from 1.4 m^{-1} for the manual data and 0 m^{-1} for the digital data (Table 5.3).

For scanline B, manual and digital measurements result in similar P10 values of 2.4 m^{-1} (manual) and 2.3 m^{-1} (digital) related to the NNE - SSW oriented fracture set. In contrast, P10 values of 0.6 m^{-1} (manual) and 0.1 m^{-1} (digital) are calculated for the W – E striking fracture set (Table 5.3).

With regard to well data, NE – SW striking fractures are most frequent, some NW – SE oriented fractures are present, and W – E orientations are mostly not detected (Figs. 5.6, 5.8, and 5.13). Comparing the dominant NE – SW fracture set density with the dominant NNE – SSW fracture set densities along scanline B (Fig. 5.13), P10 values are lower with 1.9 m^{-1}, but higher compared to that fracture set measured along scanline A (Table 5.3).

Regarding all fracture sets, the digital scanline A reveals a P10 value of 2.6 m^{-1}, the digital scanline B of 2.9 m^{-1}, and the comparable dolomite well interval of 2.7 m^{-1} (Fig. 5.13). Here, manually measured data appear higher with a P10 of 4.3 m^{-1} for scanline A and of 5 m^{-1} for scanline B (Table 5.3).

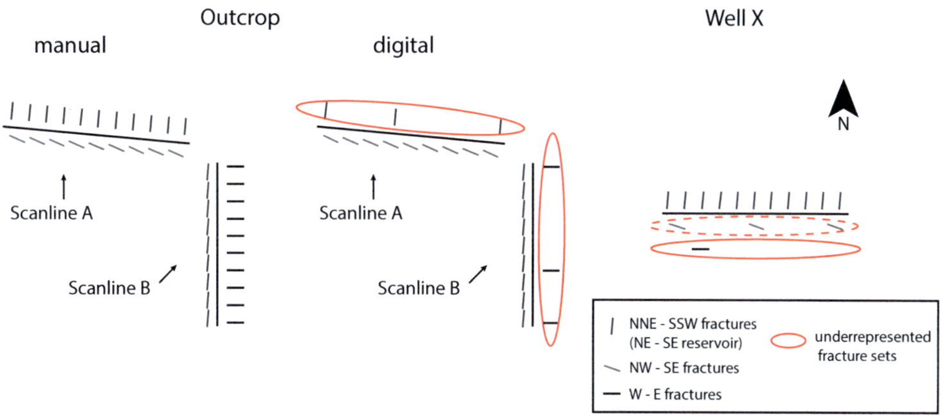

Figure 5.13. The presence of different fracture sets in the three data sets of manual outcrop, digital outcrop, and reservoir data. Fractures with almost perpendicular orientations to the scanned wall appear underrepresented in the t-LiDAR data, and fractures almost paralleling the well are underrepresented in the reservoir data.

To compare the dolomite and limestone fractures, combined fracture data of all sets are used. P10 values of all digital fracture data of the dolomites along scanline A (2.6 m^{-1}) are compared with the digital outcrop results of the limestone fractures showing a lower P10 value of 1.5 m^{-1}. In similitude, with a P10 value of 0.9 m^{-1}, the fracture spacing appears lower in the observed limestone interval than in the dolomite interval (P10 = 2.7 m^{-1}) of the well.

Table 5.3. P10 values for specific fracture sets and including all fracture data as derived from manual and digital fracture data and P10 well data for comparison. Number of fractures in brackets; *=Terzaghi-corrected values.

Fracture set		Scanline A			Scanline B		
		NW - SE*	NNE - SSW*	All	NNE - SSW*	W - E*	All
P10 [m^{-1}]	manual	1.8 (23)	1.4 (21)	4.3 (58)	2.4 (40)	0.6 (6)	5 (74)
	digital	1.8 (17)	0 (1)	2.6 (60)	2.3 (33)	0.1 (1)	2.9 (42)
		Well X					
		NW-SE*	NNE-SSW*	All	NE-SW*	W-E	All
	Well-FMI	0.9 (26)	1.9 (38)	2.7 (80)	1.9 (38)	0.3 (9)	2.7 (80)

5.6 Discussion

5.6.1 Validation of outcrop and reservoir comparability

Fracture mechanics depend on the depositional features such as composition and texture, as well as on the diagenetic history, which all affect the mechanical properties of the host rock (Gale et al., 2005; Laubach et al., 2009; Lavenu et al., 2013; Wennberg et al., 2006). Thus, the comparability of the studied outcrop and reservoir rocks needs to be validated in terms of rock type, depositional environment, and diagenesis.

Observations on outcrop carbonates and stained thin sections provide evidence that those rocks are mainly composed of mud- to wackestones with dolomitic composition (see chapter 5.5.1.1, Figs. 5.4 and 5.5). Using the neutron and density tools of the provided conventional well log data enables to define an interval of dolomitic composition and observations on thin sections from reference wells of the same reservoir match with the results of the outcrop samples. Thus, both outcrop and reservoir rocks are comparable in terms of their lithological rock type, which ultimately will reflect comparable mechanical properties.

Laminated and thinly bedded carbonates in the lower quarry parts mainly lack fossils and are intercalated with mm-thick clay layers (see Fig. 5.4). This hints to a depositional environment on a more proximal part of the Ca2 slope (cf. Clark, 1986; Huttel, 1989; Paul, 1991; Peryt et al., 2010; Reijers, 2012; Strohmenger et al., 1996; Strohmenger and Strauss, 1996). The absence of fossils in the lowermost quarry parts might even indicate the presence of a fluctuating chemocline (Paul, 1991). The top parts of this bedded unit comprise of wackestones which represents deposition on the open shelf and transition to the carbonate platform (cf. Reijers, 2012). This interpretation is supported by the presence of a massive carbonate unit with some channels of a higher energy setting overlying the bedded slope carbonates (see Fig. 5.4). Here, silica-bearing material and shadows of former ooids can be observed in thin-sections from the massive carbonates (see Figs. 5.5e and f) indicating the vicinity to a shoal environment providing e.g., silty material (cf. Reijers, 2012). Thus, the exposed carbonates represent a shallowing upwards sequence on a prograding carbonate slope from an upper slope to shoal environment. Knowledge of the reservoir including core material from former vertical wells of Field X (Koehrer et al., 2016) confirms the same setting of middle- to upper slope deposition for the subsurface.

Apart from the control of the depositional environment, diagenesis has a great influence on the rock characteristics of the Ca2 carbonates (Strohmenger et al., 1996). Diagenetic alterations can be tracked over the entire Southern Permian Basin (Clark, 1980). However, variations occur depending on the paleogeographic position in the depositional environment (Strohmenger et al., 1998). Thus, the diagenetic processes affecting those rocks need to be further compared. Detailed studies on the paragenetic sequence of the Ca2 carbonates have been published by several authors (e.g., Biehl et al., 2016; Clark, 1980; Reijers, 2012; Strohmenger et al., 1998). Here, only the main observed processes are discussed.

The first main process affecting the carbonates shortly after deposition is pervasive dolomitization (Biehl et al., 2016; Clark, 1980; Reijers, 2012; Strohmenger et al., 1998) as confirmed by observations on thin sections from outcrop samples and reference wells (see Fig. 5.5). On the former slope, this process of dolomitization was triggered by downwards migrating Mg-bearing brines from a landwards situated sabkha or saline lagoon environment indicating the beginning of evaporite deposition of the overlying A2 (Clark, 1980; Reijers, 2012). This reflux dolomitisation increased the reservoir quality of the mud-dominated slope carbonates due to the development of intercrystalline porosity (Biehl et al., 2016; Reijers, 2012).

At shallow burial, when temperatures reached a maximum of 42°C (Clark, 1980), gypsum started to convert into anhydrite releasing sulfate-rich crystal water which resulted in partial replacement of the dolomite and cementation of early formed fractures (Clark, 1980; Reijers, 2012). Intense anhydritization is present in the top parts of the quarry (see Fig. 5.4).

This beginning anhydritization is subsequently followed by a phase of calcitization (dedolomitization) (Biehl et al., 2016; Clark, 1980; Reijers, 2012) which can occur at temperatures up to 50°C (Clark, 1980) (45°C in Below, 1992; Strohmenger et al., 1998), and thus, this process is limited to a depth of approximately 680 to 1000 m (Zechstein surface temperature = 25°C, geothermal gradient = 2.5°C/100m, Clark, 1980). To replace dolomite by calcite, carbon dioxide enriched fluids and a source of calcium need to be present to ensure a high enough Ca/Mg ratio (Clark, 1980). The CO_2 was most likely released from organic matter within the carbonate slope and basin deposits (Clark, 1980; Reijers, 2012), which is also present in the lower parts of the Ührde quarry (see chapter 5.5.1.1, Fig. 5.4). Calcitization is only partially present in both, outcrop and well samples, and preferentially restricted to the replacement of former fossils or as fracture cements (see Fig. 5.5b). Thus, the amount of calcium available in the brines due to the ongoing process of anhydritization (cf. Clark, 1980) was sufficient to form calcite along open pathways, such as fractures, but not in the entire formation explaining the formation of small filled vugs present in the outcrop (see Fig. 5.5b).

The outcrop samples were later uplifted to the surface and thus, exposed to meteoric waters. However, observations do not reveal any associated cementation or dissolution processes as e.g., another phase of calcitization in the samples. Therefore, outcrop and reservoir rocks show the same type of predominant lithology, depositional environment, and diagenetic histories and thus, can be directly compared.

5.6.2 Fracture history

Overall, main NE – SW to NNE – SSW, NW – SE, and almost WNW - ESE to W – E striking fracture sets occur (see Fig. 5.6). Integrating the tectonic evolution (see chapter 5.3) with the observed fracture patterns could explain the following evolutionary models:

1. The observed WNW – ESE to W – E oriented fracture sets imply a N – S directed minimum stress direction. NNE – SSW oriented extensional tectonics from the Late Permian until the Jurassic (Betz et al., 1987; Lohr et al., 2007; Pharaoh et al., 2010; Scheck-Wenderoth et al., 2008) most likely resulted in the initiation of this fracture set (see Fig. 5.8). Additionally, normal faults and graben systems with WNW – ESE strike formed (cf. Lohr et al., 2007) resulting in corresponding fractures with moderate dip angles (cf. Nelson, 2001) (Fig. 5.14).

2. Observed N – S oriented fractures do not represent a main fracture set within the studied outcrop and reservoir data, but they are present (see Fig. 5.8). Their formation was most likely initiated by a change towards a W – E oriented extensional stress regime in the Jurassic (Betz et al., 1987; Lohr et al., 2007; Pharaoh et al., 2010; Scheck-Wenderoth et al., 2008) (Fig. 5.14).

3. Formation of the observed NW – SE fracture set required a minimum horizontal stress in NE – SW direction. Such conditions could have been met in compressive NW – SE to W – E stress regimes that were active since the Jurassic/Cretaceous boundary. Major strike-slip faults with N – S to NW – SE strike direction are reported for this regime (cf. Betz et al., 1987; Kley and Voigt, 2008; Pharaoh et al., 2010). Thus, it appears likely that the steeply dipping NW – SE fractures formed parallel to those major faults and not necessarily perpendicular to the minimum horizontal stress. Associated normal faults or graben systems with moderately dipping fractures in releasing bends could have formed simultaneously, and potential reactivation of former faults as thrusts with corresponding gently dipping fractures (cf. Nelson, 2001) perpendicular to the main stress orientation could have occurred (Fig. 5.14).

4. The main NNE – SSW to NE – SW oriented fracture sets imply WNW – ESE to NW – SE oriented minimum stress directions. They can be related to Late Cretaceous basin inversion driven by N – S to NE – SW oriented compressive stresses (Betz et al., 1987; Kley and Voigt, 2008; Lohr et al., 2007; Navabpour et al., 2017; Pharaoh et al., 2010). Those stresses resulted in transpressional reactivation structures (Pharaoh et al., 2010). Thus, NNE – SSW and NE – SW oriented steeply dipping fractures, and gently dipping thrusts and their associated fractures perpendicular to the NE – SW orientation are potentially formed (Fig. 5.14).

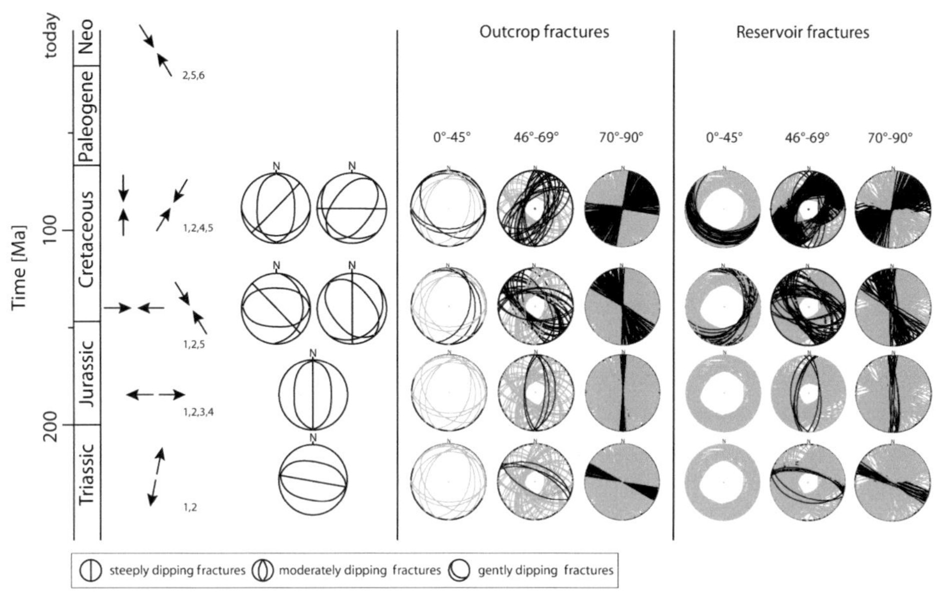

Figure 5.14. Dominant stress regimes affecting the study area since the Triassic and resulting orientations of fractures formed due to those stress fields. Corresponding fractures (black great circles) are marked in the stereographic projections of the outcrop and reservoir fractures (gray great circles) (see Fig. 5.7). 1: Lohr et al. (2007), 2: Betz et al. (1987), 3: Scheck-Wenderoth et al. (2008), 4: Pharaoh et al. (2010), 5: Kley and Voigt (2008), 6: Reiter et al. (2016).

However, the development of those fracture sets cannot be regarded as a strict relative timing of fracture formation. Results of this study also show a similar strike of steep, moderate, and gentle dip angles for each fracture set, and thus, a combination of opening and shear failure criterions influenced their formation (e.g., English and Laubach, 2017; Nelson, 2001; Olson et al., 2009). This is further highlighted by observations on abutting relationships (see Fig. 5.10), which are also indicative for a stress regime with reactivated orientations.

Focusing on fracture length, as longer fractures are older because they continued developing in newly suitable stress regimes (cf. Price and Cosgrove, 1990), NW – SE orientations appear to be older than the main NE – SW fracture set. Furthermore, the reactivation of older structures and associated new fracture opening, especially in the strike-slip regimes (cf. Betz et al., 1987; Pharaoh et al., 2010) may contribute to overprinting relationships. The similar strike of outcrop and reservoir fractures suggests that they have formed due to the same tectonic events.

In exposed rocks, further fracture formation resulting from uplift related stress release is expected (cf. Engelder, 1985; Lavenu et al., 2013). Due to ongoing erosion, the magnitude of the vertical and horizontal stress on the overconsolidated rock decreases with uplift. Close to the surface, horizontal stresses may be larger than the vertical stress which might result in the formation of bedding-parallel fractures (e.g., Mandl, 1999). Those bedding-parallel exfoliation fractures have not been measured in the field and neither been included in digital fracture detection. Additionally, poroelastic effects of the pore pressure changes and cooling effects during uplift can cause tensile horizontal stresses during uplift that might produce vertical fractures (cf. English, 2012; English and Laubach, 2017; Mandl, 1999). However, this effect strongly depends on rock characteristics. Mandl (1999) shows that such high horizontal stresses can built up in strong rocks (E-modulus > 10 GPa) if uplifted to a depth of at least a few hundreds of meters, while English and Laubach (2017) show that such uplift-related subvertical fracture formation could already be possible during an uplift from approximately 3.5 km to 3 km depth for strong (E-modulus = 62 GPa), quartz-rich sandstones. Additionally, as the actual opening of fractures requires compression of the rock material between the fractures, it is also controlled by the rock strength. Thus, only small, sub-millimeter wide fractures at several meter distance are reported to form in strong rocks due to those uplift stress conditions (Mandl, 1999, p. 176 ff.). Considering an uplift of 4 km (Kockel, 2003; Lohr et al., 2007; Reicherter et al., 2008) and using the rock characteristics measured on outcrop samples (E-modulus = 16 GPa, see Table 5.1) for the Ührde dolomites for the calculations for strong rocks given by Mandl (1999, p. 176 ff.), calculated tensile stresses are approximately 5 MPa. Perras and Diederichs (2014) report direct tensile strengths of 5.7 MPa and Brazilian tensile strengths of 8 MPa as average properties for dolomites. Their reported average UCS values of 90.4 MPa are below the measured UCS values (average 161 MPa, see Table 5.1) of the studied samples. Since an increase in UCS is generally correlated to an increase in tensile strength (Perras and Diederichs, 2014), the calculated tensile stresses of 5 MPa are likely to be less than the tensile strengths of the studied rocks. Thus, it is unlikely that uplift-related subvertical fractures opened and this aspect of additional fracturing due to uplift can be neglected in this study.

5.6.3 Validation of t-LiDAR

Supplementary to traditional field work, the use of t-LiDAR data for petroleum geological applications, as e.g., improvements in reservoir modeling, has become of increasing importance in recent years (Hodgetts, 2013).

Different approaches have been introduced to process the t-LiDAR data with the aim of achieving automated extraction of geological surface data such as bedding, faults, or fractures (e.g., García-Sellés et al., 2011; Kemeny et al., 2006; Laux and Henk, 2015;

Riquelme et al., 2014; Seers and Hodgetts, 2014; Wilson et al., 2011). One widely used method includes the generation of triangular meshes from the laser scan point cloud data (e.g., Cawood et al., 2017; Jones et al., 2009a; Kemeny et al., 2006; Seers and Hodgetts, 2014) combined with further discontinuity analysis (e.g., Kemeny et al., 2006) or moment of inertia analysis (e.g., Seers and Hodgetts, 2014) to detect, for example, fracture planes. In contrast, another approach aims at generating the surface data directly from the point cloud data (e.g., García-Sellés et al., 2011; Gigli and Casagli, 2011; Laux and Henk, 2015; Riquelme et al., 2014; Wilson et al., 2011) to avoid the additional step of mesh generation which implies large data sets and processing time. This approach incorporates an error in data accuracy due to necessary decimation steps (cf. Hodgetts, 2013). Our approach uses the point cloud data, as we are aiming at a time-efficient method of automated fracture detection. Detecting fracture planes can be done by either manual plane picking (e.g., Gaich et al., 2006; Gigli and Casagli, 2011; Gillespie et al., 2011; Sturzenegger and Stead, 2009b; Wilson et al., 2011), or applying a (semi-)automated approach based on different algorithms (e.g., García-Sellés et al., 2011; Gigli and Casagli, 2011; Laux and Henk, 2015; Riquelme et al., 2014; Wilson et al., 2011). Exemplarily, Wilson et al. (2011) introduced a workflow based on curvature calculations. However, this includes data manipulation as a regularly spaced surface is required. An advantage over other (semi-)automated approaches is, that our workflow detects fracture surfaces as ellipses along defined scanlines (see chapter 5.4, Fig. 5.3, and cf. Wüstefeld, 2018). This approach facilitates subsequent steps of fracture density analysis as a base for fracture network modeling, which is in alignment with common petroleum industry workflows for (horizontal) wells.

Results of this study indicate that t-LiDAR scanning detected all fracture orientations measured with traditional field methods (see Figs. 5.6, 5.8, and 5.12). However, a finer subdivision of the overall W – E fractures into WSW – ENE and WNW – ESE, and of the NNE – SSW into NNE – SSW and more NE – SW sets, as possible in the manual data set, cannot be done. This might indicate that input parameters of the cluster definition algorithm were chosen too widely and could be modified in further studies (e.g., lower limit for the angle between two clusters). With regard to a single scanline, it becomes evident that the question, whether a single fracture is detectable by digital means, depends on the geometry of the exposure of each fracture (cf. García-Sellés et al., 2011; Wilson et al., 2011). Fractures, whose orientations almost align the scanline orientation, have good surface exposures and thus, scan data of those areas contain a large number of points resulting in reliable orientation calculations for that fracture (see chapter 5.5.2.3, Table 5.2). The fractures that strike almost perpendicular to the scanline implies that their surfaces are not well exposed. Those fractures appear underrepresented in the digital results (see Figs. 5.12 and 5.13), as the detection algorithm requires more points to detect a surface (cf. García-Sellés et al., 2011; Wilson et al., 2011). This effect can be restricted by using different input parameters for the algorithm, e.g., by reducing the minimum amount of points needed to define one cluster (see chapter 5.4.1). Since this alters the fracture size for different scanlines, such approach was excluded. If only fracture traces are exposed, this defines the limit for this method, and other approaches (e.g., Bisdom et al., 2014; Gigli and Casagli, 2011; Seers and Hodgetts, 2016) need to be incorporated to capture more fractures. In general, it can be stated for this method that the more parallel a fracture surface is oriented to the scanned wall the higher is the ability to detect that fracture. Information about those wall-parallel fracture sets are an important advantage compared to the borehole resistivity image logs where wellbore parallel fractures are usually underrepresented (cf. Terzaghi, 1965). However, scanline-parallel fractures in outcrops are potentially biased compared to fracture sets perpendicular to the scanline (e.g., Watkins et al., 2015). Thus, if possible, different scanline orientations (scanline A and B are oriented almost perpendicular to each other in this study) should always be used to derive fracture characteristics for each fracture set.

Many studies have shown that discontinuity orientations, regardless if referring to fractures or bedding, can be displayed with a high accuracy by using terrestrial laser scan data (cf. Gigli and Casagli, 2011; Kemeny et al., 2006; Seers and Hodgetts, 2014; Sturzenegger and Stead, 2009a; Sturzenegger and Stead, 2009b; Wilson et al., 2011). The comparison of manual and digital data of this study confirm this result indicating deviations of only 5° for the dip direction and dip angle of a distinct fracture surface (see Table 5.2). In contrast, only few studies have compared further statistic parameters such as the fracture density (e.g., Gigli and Casagli, 2011; Seers and Hodgetts, 2014). The fracture intensities of this study reflect again the limited ability of the workflow to detect fracture surfaces oriented perpendicular to the scanline orientation. Thus, densities of those sets are highly underestimated and, regarding all fracture data, values only account for 60% of the manually determined values (see Table 5.3). However, for the detectable fracture sets, the digitally derived P10 values demonstrate excellent agreement with those based on manual field measurement data (see Table 5.3).

In addition to the actual fracture orientation and definition of associated fracture sets, which can be well extracted from the laser scanning data as shown above (see Figs. 5.6 and 5.8), spacing relationships can be calculated (see Table 5.3). Additional data, such as values on aperture width or filling of a fracture, cannot (yet) be assessed from point cloud data due to resolution limitations (Gigli and Casagli, 2011; Wilson et al., 2011). Thus, access to the rocks and traditional field methods are still indispensable for detailed outcrop analog studies in terms of additional data acquisition, as well as validation of the digitally derived fracture surfaces (cf. García-Sellés et al., 2011; Gigli and Casagli, 2011; Hodgetts, 2013).

5.6.4 Added values for reservoir modeling

The second scope of this study is to validate the direct transferability of the collected outcrop (manual and digital) scanline data onto the studied reservoir horizontal well data. In contrast to many previous studies using outcrop data to add lateral information to vertical borehole data (Bertotti et al., 2009; Wennberg et al., 2006), this provides the chance to compare three similar 1D scanline sets.

Fracture orientation, density distribution, size, and aperture are key input parameters for fractured reservoir models (cf. Gillespie et al., 2011; Wennberg et al., 2016).

5.6.4.1 Fracture orientation

Outcrop and subsurface fracture data are both match to the main stress states in terms of fracture orientation and fracture dip reflecting different formation processes of the tectonic history of the Southern Permian Basin (Fig. 5.14). However, mean orientations of the northerly striking fractures show NNE – SSW directions in the outcrop and NE – SW directions in the subsurface (see Figs. 5.6 and 5.8).

Local variations are mainly controlled by near-field stress perturbations in the vicinity of faults (e.g., Henk, 2005; Yale, 2003). Yale (2003) shows such influence on the near-field stress over hundreds of meters around the faults. The drilled structural compartment of the studied reservoir is bounded by faults at distances of approximately 200 m (Koehrer et al., 2016). In contrast, no faults are observed around the outcrop and the large Northern Harz Boundary Fault is at a distance of approximately 10 km (see Fig. 5.1). However, this difference could also be related to the local influence of the NNE – SSW striking 'Eichsfeld-Altmark Swell', at whose slope the outcrop carbonates were deposited (Herrmann, 1956; Paul, 1993). Thus, those different structural settings might explain the variations in the main stress orientation between the reservoir and well. However, these discrepancies are in a range that would not affect well planning as a preferential W – E orientation intersects the highest number of fractures in both settings.

5.6.4.2 Fracture density

Information on fracture densities is difficult to obtain from the subsurface (Narr, 1996; Ortega et al., 2006), as mostly vertical well data is available where the chance of underrepresenting fractures paralleling the well orientation is especially high (cf. Narr, 1996; Terzaghi, 1965). Using outcrop data helps overcome this issue (e.g., Bertotti et al., 2009; Wennberg et al., 2006). To evaluate the comparability of fracture densities of the outcrop and subsurface fractures, the same prerequisites affecting fracture occurrence, i.e. the distance to faults and their related fracture zones, need to be considered. As described before, the near-field stress regime around a fault can influence fracture orientations (cf. Henk, 2005; Yale, 2003), but also the intensity of that fracturing may vary near the fault (cf. Caine et al., 1996). Faults are characterized by their fault core and surrounded by a wider damage zone which includes fault-related smaller deformation structures as small faults or fractures (cf. Bastesen and Braathen, 2010; Caine et al., 1996; Shipton et al., 2006). As there are no faults present in the Ührde quarry or exposed nearby, outcrop fractures are expected not to be influenced by faults and their damage zones. In contrast, seismic mapping reveals a fault with approximately 150 m displacement at a distance of around 200 m to the end of the studied wellbore. Several studies have been aiming at determining a scaling relationship between the fault attributes/core and the thickness of the damage zone (e.g., Bastesen and Braathen, 2010; Bastesen et al., 2013; Childs et al., 2009; Knott et al., 1996; Shipton and Cowie, 2001; Shipton and Cowie, 2003; Shipton et al., 2006). However, composite collections of different studies show that damage zone widths depend on several factors such as lithology, porosity, or fault linkage (cf. Shipton and Cowie, 2003), and thus, result in a wide range of possible width ratios (e.g., Shipton et al., 2006; Torabi and Berg, 2011). Following the correlation by Shipton and Cowie (2001) of porous rocks, where the damage zone width equals 2.5 * fault throw, the subsurface fault of this study would have a damage zone of 333.5 m (166.75 m on each side of the fault), and thus, should not interfere with the fracture density along the well. Considering that damage zone widths are even smaller in tight rocks (cf. Bastesen and Braathen, 2010; Shipton and Cowie, 2003), the actual damage zone might be significantly lower. If referring to power law relationships (e.g., damage zone thickness = 0.019 * fault displacement^0.8, Shipton et al., 2006) between the fault displacement and damage zone thickness generated from multiple global data sets (cf. Childs et al., 2009; Shipton et al., 2006; Torabi and Berg, 2011), the determined damage zone thickness of only about 1 m would be even lower. Therefore, no fault-related influence on the fracture density distribution is assumed for the outcrop analog or the subsurface, and thus, the results of both can be compared with each other.

Another general controlling factor on the fracture density is correlated to the mechanical rock properties (Laubach et al., 2009; Lavenu et al., 2013). This includes the rock strength evolution related to lithological properties and diagenetic overprints (cf. Gale et al., 2005; Laubach et al., 2009; Lavenu et al., 2013; Wennberg et al., 2006), which are shown to be comparable for the outcrop and reservoir dolomite.

Although the influence of fracture formation due to faulting is shown to be negligible, and outcrop and reservoir rocks feature the same rock mechanical properties, the comparison of the determined P10 values of this study shows differences between the outcrop and subsurface data (see Fig. 5.13, Table 5.3).

With regard to the NE – SW (well) to NNE – SSW (outcrop) fracture set, which represents the main orientation in both, outcrop and subsurface, the effect of systematically underrepresented fractures oriented parallel to the well (cf. Terzaghi, 1965) is assumed to be negligible, as this set is oriented at a high angle to the well orientation (see Fig. 5.13, Table 5.3). Here, using the average determined from the P10 values of both scanline orientations from the outcrop would exactly match the P10 value obtained from well data.

However, if using quarries as analogs, local variations in fracture characteristics can occur due to production processes, and thus, determined parameters should not be based on one single scanline to account for such potential variations. Furthermore, the effect of stress on fracturing needs to be taken into account (e.g., Ferrill et al., 1999; Wilson et al., 2011). Though stress calculations suggest no significant additional fracturing due to uplift, fractures with vertical persistence of a few centimeters (see Fig. 5.7b) in the outcrop might contribute to higher densities but are not relevant for the reservoir fracture networks. In addition, due to the relative orientation of reservoir fractures to the present-day stress field, long, persistent fractures with large apertures in the outcrop might appear almost closed in the subsurface. Thus, future studies should further deal with these effects. With regard to t-LiDAR derived data, a systematic bias can occur depending on the fracture geometry as discussed above, and thus, careful data evaluation including validation with manual measured data is needed to avoid over- or underestimated flow performance in the reservoir models (cf. Seers and Hodgetts, 2014).

Using the P10 for data comparison in this study, and thus, reducing the 3D point cloud data set to one dimension might seem a drawback at first sight, but the scope of this study is the comparison with equal data sets from manual scanlines and horizontal well data. Furthermore, P32 values (fracture area per unit volume, Dershowitz and Herda, 1992) can be derived by upscaling P10 values as shown by other authors (Bisdom et al., 2014; Sanderson and Nixon, 2015; Zeeb et al., 2013). For Terzaghi corrected analyses, P10 values equal the P32 values for a fracture set (Sanderson and Nixon, 2015), and thus, values given in this study for the different fracture sets could be used as input for a first reservoir model. Another approach is shown by the recent work of Seers et al. (2017), who directly obtained modeled fracture porosity characteristics from digital outcrop models as input for reservoir models.

As the t-LiDAR workflow has been successfully validated, data acquisition has been widened to an inaccessible quarry part, where limestone is exposed. Detection of fracture planes along a scanline comparable to the dolomite scanlines allows to evaluate potential fracture density relationships to the type of lithology. In similitude, a limestone interval from the horizontal well has been additionally analyzed. Results of both, outcrop and reservoir indicate lower densities in the limestone intervals than in the dolomite which fits to expected lower fracturing of limestone than dolomite (e.g., Ortega et al., 2006).

5.6.4.3 Fracture size and aperture

Limited dimensions of rock exposures generally reduce the value of outcrops for fracture length quantifications (cf. Ortega et al., 2006; Zeeb et al., 2013). However, persistence measurements provide insight on reservoir-relevant fractures, which exceed bed thickness (see Fig. 5.7b). Together with observations of bitumen-impregnated fracture surfaces (see Fig. 5.9b) representing such large persistent fractures, this provides evidence that the studied outcrop fractures are the same that can be found in the reservoir. Aperture measurements from outcrops are often regarded as added values for reservoir modeling, but to determine actual reservoir fracture apertures at depth, the present-day stress field needs to be taken into account (cf. Ferrill et al., 1999; Wilson et al., 2011). Another aspect that needs to be considered is the process of karstification in carbonate rocks. Outcrop fractures act as pathways for meteoric waters, and thus, karst dissolution would be most effective along those flow paths. Therefore, measured outcrop fracture apertures might contain additional widening due to dissolution (cf. Dreybrodt, 1990), and using those values would overestimate actual aperture widths at depth. However, dissolution rates of dolomite are significantly lower than of calcite, resulting in a much more limited influence of karst processes in dolomite rocks (Krawczyk and Ford, 2006; Liu et al., 2005). As a result, aperture values obtained from the outcrop (see Fig. 5.7a) are only suggested to be used as input for reservoir models for rough estimations.

Those two parameters, fracture size and aperture, are especially important for correct flow simulations (cf. Seers and Hodgetts, 2014; Zeeb et al., 2013), however, determining exact numbers remains challenging as manual outcrop data only provide approximate numbers and extractions from point cloud data are not yet possible due to resolution limits (cf. Gigli and Casagli, 2011; Wilson et al., 2011).

As reservoir fracture data cannot provide information about the linkage of fractures, the comparison with outcrop analog data is essential to model the correct fracture network responsible for fluid flow (cf. Zeeb et al., 2013). The results of our study can be used to improve fracture network understanding as abutting relationships were mapped indicating good connections between all main fracture sets (see Fig. 5.10). To get this information fracture traces were mapped on stitched photographs from the quarry floor as it is not yet possible to detect them in the point cloud data. However, reservoir fractures were formed due to similar events which is why an equally well connected fracture network in the subsurface is assumed, but again, actual flow paths are mainly controlled by the present-day stress field (cf. Ferrill et al., 1999; Wilson et al., 2011).

In general, the comparison of manual and digital outcrop scanline data with horizontal well data indicates that fracture orientation, density, and size could be used as input parameters for actual reservoir modeling and well planning in analogous subsurface fields. Though t-LiDAR scanning provides abundant advantages regarding large amounts of data, additional data of inaccessible areas, and quick data collection (e.g., Hodgetts, 2013), results need further validation with manual data as exemplarily distinct fracture sets occur underrepresented depending on exposure characteristics.

5.7 Conclusions

We successfully validated the semi-automated workflow based on t-LiDAR scanning which allows to generate large amounts of data even for inaccessible areas. The two scanlines of this study and the input parameters chosen for this setting indicate deviations for the dip direction and dip angle of the fracture planes of average 5° between the digitally and manually derived orientations. A bias occurs in the digital dataset for orientations almost perpendicular to the outcrop wall, which are only limitedly exposed and thus, appear underrepresented.

The Upper Permian Zechstein Ca2 carbonates of the Ührde quarry SW of the Harz mountains can be used as a direct analog for gas-producing Ca2 reservoirs in the North German Basin approximately 130 km to the northwest, as they comprise similar lithology and rock texture of dolo-mudstones to dolo-wackestones, similar depositional environments of a carbonate platform slope setting of the former Southern Permian Zechstein Basin, comparable diagenetic overprint with stages of dolomitization, dedolomitization, and redolomitization, as well as tectonic fracturing events.

Fracture characteristics from the outcrop and subsurface are shown to be comparable due to similar rock mechanics, negligible processes of uplift related fracturing, and the observation of bitumen-impregnated fracture surfaces on persistent fractures in the outcrop. Both, outcrop and reservoir fractures reveal a good comparability in terms of dominant northerly strike orientations suggesting preferentially W – E well orientations, though true mean orientations show slight deviations. Although all fracture orientations are observed in the outcrop and reservoir data, main strike orientations vary which might be due to the influence of near-field stress perturbations around faults bounding the reservoir and the position of the outcrop on a NNE – SSW striking paleo-high.

Fracture intensities of t-LiDAR data from the outcrop and reservoir fractures match extremely well (t-LiDAR: 2.6 m^{-1}; well: 2.7 m^{-1}), indicating a comparable bias due to an underrepresentation of fractures paralleling the well and in similitude, fractures perpendicular to the wall orientation are under-detected by laser scanning. Fracture sizes needed for DFN reservoir modeling can only be obtained from outcrops and fracture networks can be defined on fracture maps taken from pavements. Results of this study show a strong need for manual validation of t-LiDAR data, but also a great potential to incorporate that data in reservoir models to reduce uncertainties.

6 Present-day stress control on fluid migration pathways: Case study fractured Zechstein carbonates, NW-Germany

6.1 Abstract

Understanding the orientation and connectivity of fracture systems in tight reservoirs is essential to reduce uncertainties in reservoir development and production. However, further knowledge of the actual flow potential of the fractures can be controlled by the present-day stress field. Dilation- and slip tendency analyses can improve fracture permeability estimations and thus, aid estimating the hydrocarbon recovery of a field.

This study focuses on fracture data of a naturally-fractured, relatively tight (matrix permeabilities of 0.01 to 1 mD) and gas-producing Late Permian (Zechstein) reservoir in the Southern Permian Basin, northern Germany. Fracture data are obtained from resistivity image log data of a 775 m long horizontal well, and show principal NE – SW orientation. To quantify the potential of those fractures to contribute to the flow within the reservoir, dilation (Td)- and slip (Ts) tendencies are calculated. Two different stress scenarios are defined based on known variations in the orientation of the horizontal stresses between formations below and above the Zechstein salt in an offset field approximately 5 km away.

Results of Td and Ts strongly depend on the strike and dip angle of the fractures, with sub-vertical fractures showing highest potential to dilate while fractures with dip angles of 60° are favorably oriented for slip. Fractures with orientations paralleling the orientation of the principal horizontal stress have highest slip- and dilation tendencies. Adding that information to the calculation of fracture permeability results in a significant reduction of the flow potential due to the influence of the stress field. Thus, incorporating stress field data helps reducing risks in field development planning.

6.2 Introduction

Fractures are known to significantly control hydrocarbon flow paths in many reservoirs, and thus, play an important role in terms of reservoir characterization and planning of well locations and -trajectories. Therefore, several studies focused on an improved understanding on various fracture attributes (cf. Gale et al., 2005; Gillespie et al., 2011; Lamarche et al., 2012; Lavenu et al., 2013; Narr, 1996; Peacock et al., 2003; Wennberg et al., 2006; Wilson et al., 2011; Zeeb et al., 2013). Fewer studies used such information to link it to reservoir production data (e.g., Hennings et al., 2012; Narr, 1991; Solano et al., 2011) or to actually derive the optimum drilling direction (e.g., Nolen-Hoeksema and Howard, 1987). Especially carbonate reservoirs often occur naturally fractured (Garland et al., 2012; Lamarche et al., 2012; Nelson, 2001). In addition to gaining knowledge about the occurrence of fractures in subsurface reservoirs and their network geometries, which could be estimated by studying suitable outcrop analogs (cf. Becker et al., 2018), it is essential to understand the hydraulic conductivity potential of those fractures. Depending on their orientations, mineral fill, connectivity, and size distribution (e.g., Hilgers et al., 2006; Laubach, 2003; Long and Witherspoon, 1985; Philip et al., 2005), opening-mode fracture arrays and faults can have a wide range of impacts on fluid flow. Although many fractures are relatively stiff and insensitive to current stress state (e.g., Laubach et al., 2016), for compliant fractures and faults that are critically stressed, orientation and present-day in situ stress field can have a strong impact on fluid flow (cf. Barton et al., 1995; Ferrill et al., 1999; Hennings et al., 2012; Rogers, 2003; Wiprut and Zoback, 2002; Zeng et al., 2015).

To understand the critical stress state of a randomly oriented fracture surface within a stress field, slip and dilation tendency analyses provide a useful tool to quantify its reactivation potential (e.g., Barton et al., 1995; Collettini and Trippetta, 2007; Ferrill et al., 1999; Lisle and Srivastava, 2004; Morris et al., 1996; Morris and Ferrill, 2009; Neves et al., 2009). The dilation tendency addresses the likelihood of a fracture to act as open fluid pathways in a defined stress field, i.e. by enlarging the fracture aperture (cf. Ferrill et al., 1999). The slip tendency describes the fracture's potential to overcome the frictional resistance to sliding (e.g., Ferrill et al., 1999; Morris et al., 1996). Considering an anisotropic stress field ($\sigma_1 > \sigma_2 > \sigma_3$), as is generally the case for sedimentary basin hosted reservoirs, better flow characteristics occur if fractures with high dilation- and slip tendencies are present (cf. Barton et al., 1995; Collettini and Trippetta, 2007; Ferrill et al., 1999; Morris et al., 1996; Morris and Ferrill, 2009; Rogers, 2003; Sibson, 2000). A better knowledge of the leaking or sealing characteristics of faults and fractures can highly contribute to successful production performance. Hence, the applicability of this technique either for understanding past slip characteristics or predict future behavior has been shown in several studies in the last years, not only for hydrocarbon reservoirs but also with regard to geothermal exploration, CO_2- and nuclear waste storage (e.g., Barton et al., 1995; Collettini and Trippetta, 2007; Ferrill and Morris, 2003; Lisle and Srivastava, 2004; Meixner et al., 2016; Moeck et al., 2009; Sathar et al., 2012; Sibson, 2000; Wiprut and Zoback, 2002).

Ideal dilation and slip tendency analyses require the knowledge of the orientations and magnitudes of the three principal stresses, knowledge of fluid pressure, and friction and cohesion characteristics of the studied rocks (Morris and Ferrill, 2009). In the case of northern Germany, a long history of hydrocarbon production provides stress data of well above 1000 wells for various target horizons, published e.g., by Reiter et al. (2016), Fleckenstein et al. (2004), Röckel and Lempp (2003), Roth and Fleckenstein (2001). The published North German stress data not only show a regional W – E trend in main horizontal stress orientation from NW – SE towards NE – SW (Fleckenstein et al., 2004; Reiter et al., 2016; Roth and Fleckenstein, 2001), but they also appear decoupled vertically due to the occurrence of the Zechstein salt resulting in different sub- and supra-salt stress regimes (Fleckenstein et al., 2004; Röckel and Lempp, 2003; Roth and Fleckenstein, 2001; Scheck and Bayer, 1999). Although those data are summarized, little has been published on the influence of the in-situ stresses on reservoir fracture or fault behavior so far. Exemplarily, Moeck et al. (2009) use slip tendency analysis to characterize the critical stress states of Rotliegend sandstones and volcanics of a geothermal reservoir in NW Germany.

We apply dilation and slip tendency on fractures along a horizontal well through a Zechstein Ca2 carbonate reservoir (Fig. 1) (cf. Well X in Becker et al., 2018). As nearby field data hint towards vertical variations in principal stress orientations in the sub- and supra-salt parts (Fleckenstein et al., 2004) (Fig. 2), we define two different stress scenarios for the calculations. Results are compared and discussed with regard to their influence on reservoir fracture permeability.

6.3 Geological setting and tectonic regime

The studied reservoir data is obtained from a horizontal development well of a producing naturally-fractured gas field in Lower Saxony, Germany (cf. Field X in Koehrer et al., 2016). It is situated approximately 60 km south of Bremen, Germany (Fig. 6.1). Structurally, this area belongs to the Lower Saxony Basin, which was formed as a sub-basin within the Southern Permian Basin since the Late Permian (Betz et al., 1987; Binot et al., 1993).

Those carbonates represent part of the second of a total of seven Zechstein depositional cycles (Z1-Z7) in northern Germany (cf. Peryt et al., 2010; Strohmenger et al., 1998). While the older three cycles feature carbonate-evaporite sequences, the younger cycles lack the deposition of carbonates (Peryt et al., 2010; Strohmenger et al., 1998). The largest number of producing hydrocarbon fields occurs in the Ca2 (Stassfurt) carbonate formation of the second cycle with production from the North Sea to eastern Poland (Fig. 6.1a) (LBEG, 2016; Peryt et al., 2010). As evaporitic depositions (A2, Basal anhydrite and younger salt sequences) were deposited on top of the Ca2 carbonates, they act as a basin-wide top seal (Karnin et al., 1996; Peryt et al., 2010; Sannemann et al., 1978; Strohmenger et al., 1996; Strohmenger and Strauss, 1996). Carboniferous coal measures are identified as main source for the Ca2 reservoirs and in some parts of the basin, organic-rich Ca2 carbonates contribute to the hydrocarbon amount (Peryt et al., 2010; Strohmenger et al., 1996).

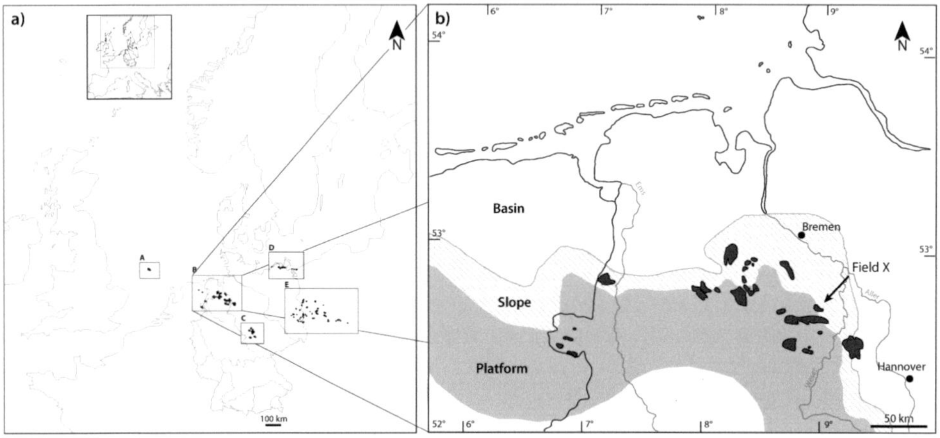

Figure 6.1. a) Distribution map of Zechstein Ca2 hydrocarbon-bearing reservoirs of the Southern Permian Basin, A = North Sea offshore; B = NW Germany and the Netherlands; C = Thuringian Basin; D = North Germany; and E = Poland and NE Germany (modified after LBEG, 2016; Peryt et al., 2010). b) Facies map showing the location of the studied Field X and adjacent NW German Ca2 fields within the former Southern Permian Basin slope to platform environment (modified after LBEG, 2016; Peryt et al., 2010; Pharaoh et al., 2010; Strohmenger et al., 1996).

In NW Germany, Ca2 carbonates were deposited along the southern margin of the Southern Permian Basin (Fig. 6.1b). Their thicknesses generally range between 20 to 80 m on the platform, reach up to 250 m on the slope, and do not exceed 10 m in the basin facies (Strohmenger et al., 1996; Strohmenger and Strauss, 1996). Regarding the field in the focus of this study (Field X in Fig. 6.1b), Ca2 slope facies comprise thicknesses of up to 250 m with successful sour gas production since 1972 (Koehrer et al., 2016).

Natural fractures enable economic production from those otherwise tight carbonate reservoirs. Fracture formation started during early burial due to an extensional post-Variscan regime that had been active since the Late Permian and Triassic (Betz et al., 1987; Glennie and Underhill, 1998; Lohr et al., 2007). Subsequently, crustal separation due to the opening of the Tethys ocean resulted in a rotation of the NW European stress field from a NNE – SSW to a W – E oriented extensional regime (Betz et al., 1987; Franzke et al., 2007; Lohr et al., 2007; Pharaoh et al., 2010; Scheck-Wenderoth et al., 2008). A NW-trending transtensional regime was active at the Upper Jurassic/Lower Cretaceous boundary (Betz et al., 1987; Kley and Voigt, 2008; Pharaoh et al., 2010) resulting in paralleling faults and associated fractures. Since the Upper Cretaceous, northern Germany was dominated by N – S to NE – SW oriented compression due to the Africa-Iberia-Europe convergence, which initiated basin inversion in the Upper Cretaceous (Betz et al., 1987; Kley and Voigt, 2008; Lohr et al., 2007; Pharaoh et al., 2010). This event was accompanied by an uplift of 1-2 km in the area of the studied field (Lohr et al., 2007) and by fault reactivation as thrusts and strike-slip (Navabpour et al., 2017) and associated fractures. Today's stress regime has been present since the late Oligocene, where a counterclockwise rotation of NE – SW to NW – SE Alpine compression started (Betz et al., 1987; Fleckenstein et al., 2004; Glennie and Underhill, 1998; Kley and Voigt, 2008; Reiter et al., 2016). Those orientations are well displayed in maximum horizontal stress orientation data obtained from sub-salt Rotliegend and Carboniferous wells showing NW – SE directions in NW Germany (Fig. 6.2) and rotating towards NE – SW directions in the eastern part of Germany (Fleckenstein et al., 2004; Reiter et al., 2016; Roth and Fleckenstein, 2001). In contrast, maximum horizontal stress data from supra-salt formations indicate higher variations in orientation (Fleckenstein et al., 2004) (Fig. 6.2), and thus, hint to an influence of the Zechstein salt acting as a decoupling horizon for the stress (Fleckenstein et al., 2004; Röckel and Lempp, 2003; Roth and Fleckenstein, 2001; Scheck and Bayer, 1999). The relative stress magnitudes can be inferred from earthquake focal mechanisms. Northern Germany is characterized by low tectonic seismicity, even historical catalogues show only few events (Grünthal, 2004; Grünthal et al., 1998; Reicherter et al., 2008). In recent years, some fault reactivation has been observed. From the focal mechanism solutions of the events in Northern Germany, a normal faulting regime with the maximum stress being oriented vertically can be derived (e.g., Fleckenstein et al., 2004; Röckel and Lempp, 2003).

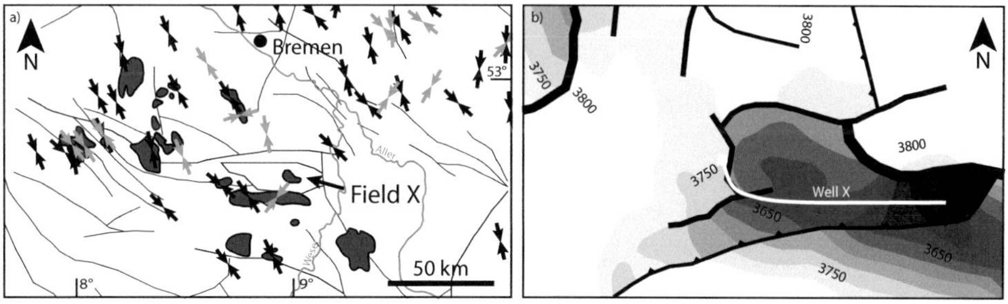

Figure 6.2. a) Distribution of the maximum horizontal stresses around the NW German Ca2 fields, together with the orientation of main faults, showing general NW – SE orientations in sub-salt (black arrows) and scattering towards NE – SW orientations in supra-salt (gray arrows) formations (stress data from Fleckenstein et al., 2004; fault data from Karnin et al., 2006; locations of Zechstein fields from LBEG, 2016). b) Top Ca2 map of the part of the studied field with the location of the studied horizontal well.

6.4 Materials and methods

Subsurface reservoir data was provided from a W – E oriented, 775 m long horizontal well section at a depth of approximately 3.7 km (Fig. 6.3). Borehole resistivity image log data (FMI tool from Schlumberger©) comprises of 1050 open fractures and 98 veins. A combination of resistivity-, density-, and photoelectric factor image log data was used to determine the fracture types. High contrasts in the resistivity image log indicate open fractures and changes in the density- and photoelectric factor image logs show differences in the mineralogy that indicate the presence of cemented fractures, i.e. veins.

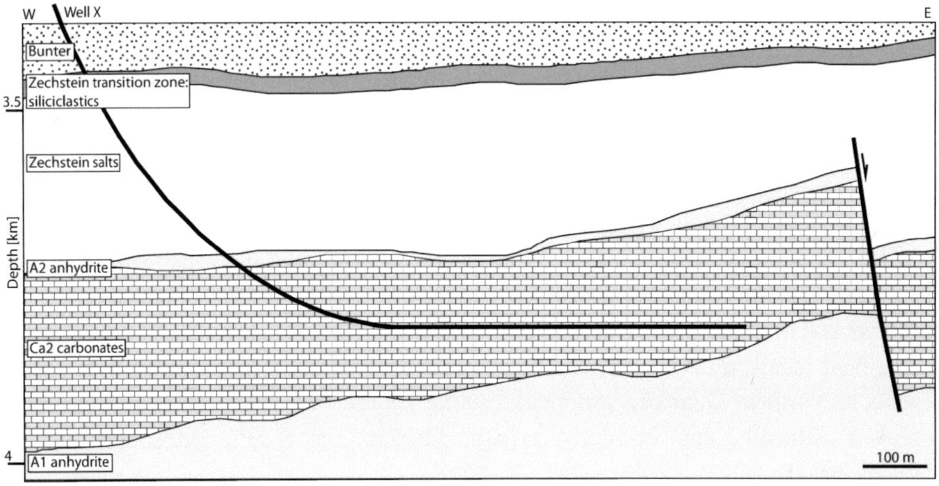

Figure 6.3. Schematic sketch of the well path of Well X and subsurface reservoir setting of the field X Ca2 carbonates sealed by A2 anhydrites and younger Zechstein salts (Z2 and younger) on top.

6.4.1 Dilation tendency and slip tendency

In addition to fracture orientations and densities of the reservoir fractures (cf. Becker et al., 2018; Becker et al., 2017b), the understanding of the reactivation potential of each fracture within the present day stress field is essential. This reactivation potential can be inferred from analyses of the slip and dilation tendency of the fractures.

Both, dilation and slip tendency need the information about the present stress field to derive a stress tensor for the reservoir depth. Stress information indicate a normal faulting regime (Fig. 6.2) as generally assumed to be present in northern Germany (cf. Fleckenstein et al.,

2004; Röckel and Lempp, 2003), where the maximum principal stress σ_1 ($=\sigma_v$) is oriented vertical and σ_2 ($=\sigma_H$) and σ_3 ($=\sigma_h$) are horizontal (e.g., Morris and Ferrill, 2009).

Two different scenarios were defined to represent the variations in the orientation of the maximum horizontal stress in sub- and supra-salt formations. In the neighboring field of the studied Field X, those orientations differ by 90° (Fig. 6.2) as shown by Fleckenstein et al. (2004). Scenario 1 was set to represent the sub-salt stress orientations of NW – SE for σ_2 ($=\sigma_H$) and correspondingly, of NE – SW for σ_3 ($=\sigma_h$). Scenario 2 reflects the supra-salt horizontal stress orientations with NE – SW orientation for σ_2 ($=\sigma_H$) and NW – SE orientation for σ_3 ($=\sigma_h$).

For both scenarios, σ_1 was calculated from the overburden (e.g., Secor, 1965) to define the stress tensor at certain depths (Table 6.1), and σ_3 was determined using literature values (Röckel and Lempp, 2003). For σ_2, only slightly higher magnitudes were assumed as the differing states of σH and σh in the Northern German Basin (cf. Reiter et al., 2016) are indicative of similar magnitudes of σ_2 and σ_3. 'Effective' stresses σeff control the actual reactivation potential of rock surfaces, i.e. faults or fractures, in the subsurface. They can be derived from the total stress magnitudes reduced by the amount of pore fluid pressure (e.g., Secor, 1965). The pore fluid pressure in the area of investigation appears to be overpressured by about 20% with respect to the hydrostatic pressure (Table 6.1) and was determined using actual data of the studied field published by Holloway et al. (1996) (p_f=38.1 MPa at 3150 m depth), which are in accordance with fluid pressure data shown by Karnin et al. (2006).

Table 6.1. Stress tensor used for scenario 1 and 2 including the magnitudes given in MPa of the principal stresses σ_1 ($=\sigma_v$), σ_2 ($=\sigma_H$), and σ_3 ($=\sigma_h$), together with the hydrostatic pressure (p_f) at different burial depths;*values after Röckel and Lempp (2003).

Depth	$\sigma_1 = \sigma_v$	$\sigma_2 = \sigma_H$	$\sigma_3 = \sigma_h$*	p_f
[km]	[MPa]	[MPa]	[MPa]	[MPa]
1	26	18	16.3	12
2	52	36	32.6	24
3	78	54	48.9	36
4	104	72	65.2	48

The dilation tendency describes the potential of a fracture to dilate in a given stress field, and thus, its likelihood to contribute to the flow within the reservoir. The normal stress acting on the fracture surface controls this ability of a fracture to dilate, i.e. the aperture of that fracture (cf. Ferrill et al., 1999). The dilation tendency is defined as the difference between the effective normal stress $\sigma_{n,eff}$ and $\sigma_{1,eff}$ normalized by the differential stress ($\sigma_{1,eff} - \sigma_{3,eff}$) (Ferrill et al., 1999):

$$Td = \frac{(\sigma_{1,eff} - \sigma_{n,eff})}{(\sigma_{1,eff} - \sigma_{3,eff})}$$

resulting dilation tendencies are in the range of 0 to 1, representing an upper limit scenario ($\sigma_{n,eff} = \sigma_{3,eff}$) and a lower limit scenario ($\sigma_{n,eff} = \sigma_{1,eff}$) (Fig. 6.4a).

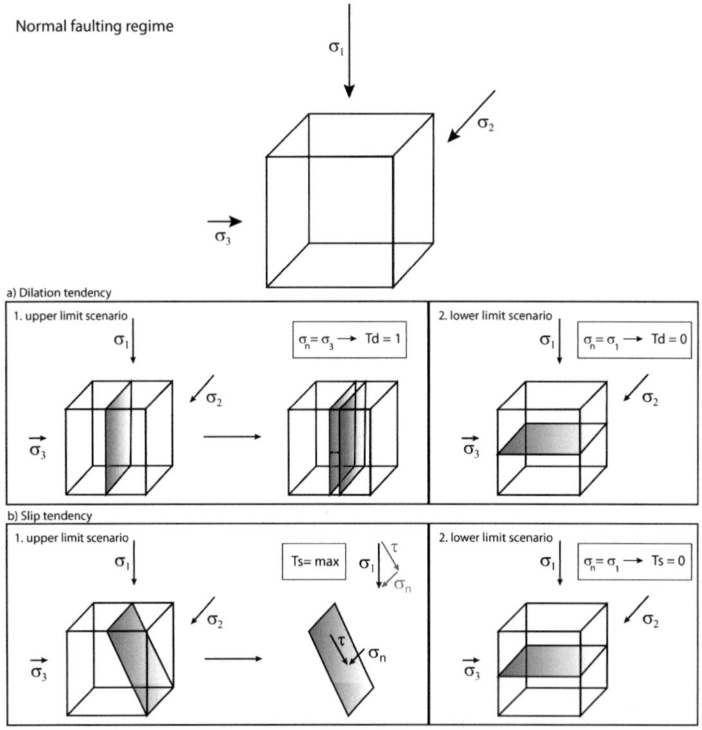

Figure 6.4. Scenarios for the maximum and minimum dilation- and slip tendencies in a normal faulting regime representing the principal stress distribution active in the study area, which implies that σ_1 is oriented vertical (per definition σ_1 (=σ_v) > σ_2 (=σ_H) > σ_3 (=σ_h)). a) Dilation tendency: For fracture planes oriented parallel to $\sigma_1 - \sigma_2$, maximum dilation can be observed. In contrast, minimum dilation occurs on fractures oriented in $\sigma_2 - \sigma_3$ orientation. b) Slip tendency: Maximum slip occurs on fractures with $\sigma_1 - \sigma_2$ orientation, while no slip can occur on fractures oriented parallel to the $\sigma_2 - \sigma_3$ plane.

Depending on the cohesion and friction coefficient, a fracture plane might slip in the present stress field (cf. Ferrill et al., 1999; Neves et al., 2009). Regarding a pre-existing plane of weakness, i.e. a fracture surface in this case, the influence of cohesion can be neglected for 'reshearing' along that plane (cf. Morris et al., 1996; Neves et al., 2009; Sibson, 2000). If the shear stress to normal stress ratio exceeds the friction coefficient μ, slip on a fracture can occur. This ratio is addressed as the slip tendency (cf. Byerlee, 1978; Morris et al., 1996; Neves et al., 2009) (Fig. 6.4b):

$$Ts\ (\triangleq \mu) = \frac{\tau}{\sigma_{n,eff}}$$

In order to achieve good comparability with the dilation tendency, slip tendency values were normalized by its maximum value to show the same range (Ts' = 0 to 1) (cf. Neves et al., 2009).

The dilation and slip tendency were calculated for the open and closed fracture data of the horizontal well using the software 3Dstress® based on the determined stress tensor each, for scenario 1 and 2. Fracture strike and dip angle data, and their position at depth were needed as input parameters.

6.5 Results

6.5.1 Subsurface fracture data

Orientations of the 1050 open fractures obtained from resistivity image log data show a main NE – SW strike direction. Minor fracture sets strike NW – SE and NNE – SSW (Fig. 6.5a). Plotting the fracture data as poles (Fig. 6.5b) indicates additional information about the dip angle of the fractures as shown by Becker et al. (2018). For 50% of the fractures, a steep dip angle (70-90°) can be observed, while moderately (46-69°) dipping fractures account for 30% of the total open fractures, which leaves 20% of the fractures with dip angles <45°.

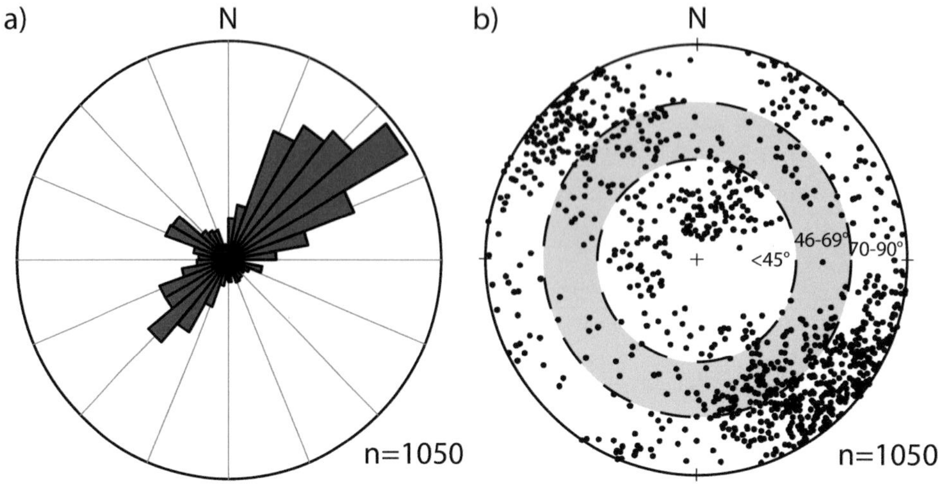

Figure 6.5. Open fracture data indicating main fracture strike orientations of NE – SW, plotted as a) rose diagrams (strike) and b) poles in the lower hemisphere of a stereographic projection. Dip angle classes are shown by gray shading in b).

6.5.2 Dilation- and slip tendency results for scenario 1

Scenario 1 represents the sub-salt stress orientations of NW – SE orientation for σ_2 (=σ_H). Regarding the open fracture data along the horizontal well, maximum calculated dilation (Td) and normalized slip (Ts') tendencies of 1 occur for fractures with orientations parallel to the orientation of today's maximum horizontal stress σ_H. For fractures oriented perpendicular to σ_H, maximum values of Td are 0.81 and a normalized Ts' of 0.6 occurs. However, wide ranges in Td and Ts' values occur disregarding the strike orientation. With regard to the dip angle, fractures show highest Td values between 0.8 and 1 for dip angles of 90°. In contrast, highest Ts' values are calculated for fractures with dip angles of 60° and for angles of 90°, lowest Ts' values are present (Fig. 6.6).

For this calculated stress scenario, maximum normalized slip tendencies correspond to calculated (not normalized) slip tendencies of 0.62.

Observations on the open fracture data show that both, the fracture strike orientation within the present stress field and the dip angle of a fracture influence its dilation- and slip tendencies. In general, it shows that the steeper the dip angle the higher is the potential of that fracture to dilate in the stress field. In addition, the potential of a fracture to slip is highest for moderately dipping fractures.

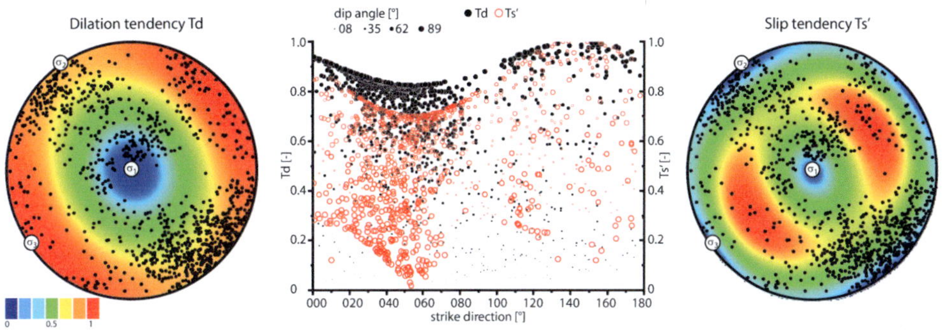

Figure 6.6. Results for scenario 1. Left figure: Reservoir fracture orientations plotted as dip direction/dip with color-coding according to the dilation tendency in a stereographic projection. Central figure: Diagram showing dilation and slip tendency as a function of the fracture strike direction for different dip angle categories. Right figure: Reservoir fracture orientations plotted as dip direction/dip with color-coding according to the slip tendency in a stereographic projection. Red colors indicate maximum Td and Ts' values.

6.5.3 Dilation- and slip tendency results for scenario 2

Scenario 2 represents the supra-salt stress orientations of NE – SW orientation for σ_2 (=σ_H). For scenario 2, calculations of the dilation and slip tendencies on the open fractures along the well show maximum values for NE –SW strike orientations reflecting the orientation of the maximum horizontal stress. In contrast, values only reach Td values of 0.77 and normalized Ts' values of 0.7 for fractures striking perpendicular to σ_H. Similar to scenario 1, wide ranges in Td and Ts' values were calculated for all measured orientations. With regard to the dip angle, a maximum dilation tendency is reached for a dip angle of 90°, and maximum slip tendencies occur for moderate angles of 60° (Fig. 6.7). Those observations match those for scenario 1, but appear even more pronounced with smaller variations.

Maximum normalized slip tendency values are equivalent to calculated (not normalized) values of 0.63 for the defined stress regime.

Similar to the observations of scenario 1, the results of scenario 2 indicate control of both, the strike direction of each fracture and the fracture dip angle. While steepest fractures have the highest potential to dilate in the stress field, the highest potential to slip is reached for moderately dipping fractures.

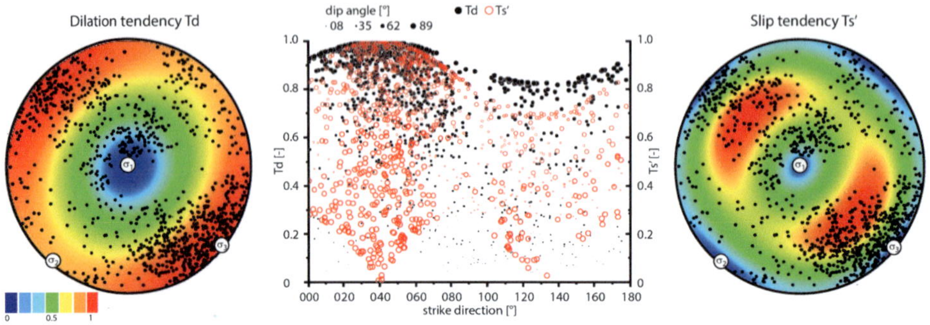

Figure 6.7. Results for scenario 2. Left figure: Reservoir fracture orientations plotted as dip direction/dip with color-coding according to the dilation tendency in a stereographic projection. Central figure: Diagram showing dilation and slip tendency as a function of the fracture strike direction for different dip angle categories. Right figure: Reservoir fracture orientations plotted as dip direction/dip with color-coding according to the slip tendency in a stereographic projection. Red colors indicate maximum Td and Ts' values.

6.6 Discussion

6.6.1 Controlling factors on Td and Ts

6.6.1.1 Fracture orientation and dip angle vs Td

Regarding the opening of a fracture in a normal faulting regime, maximum opening would be expected for fractures resembling the $\sigma_v - \sigma_H$ ($\sigma_1 - \sigma_2$) plane (see Fig. 6.4), i.e. steep fractures oriented parallel to σ_H (cf. Ferrill et al., 1999). In accordance to that theory, dilation tendency analyses of this study show that fractures with orientations parallel to the maximum horizontal stress σ_H feature highest values for both stress scenarios. In similitude, fractures oriented perpendicular to σ_H show minimum potential to dilate. However, with regard to the proportion of fractures showing high dilation tendencies $\geq$0.8 (cf. Ferrill et al., 1999), both scenarios reveal variable results. For scenario 1, 390 reservoir fractures have Td values higher than 0.8, corresponding to a proportion of 37% of all calculated fractures. In contrast, such high values were determined for around 53% (552) of all fractures in scenario 2. As scenario 1 and 2 solely differ in the orientation of σ_H, those results indicate the strong influence of the fracture surface orientation relative to the maximum stress orientation (e.g., Morris et al., 1996). In both scenarios, σ_1 is oriented vertical, and thus, variations in slip tendency depend on the orientation of σ_2 as σ_H. As the mean strike orientation of all fractures is approximately NE – SW (see Fig. 6.5), which aligns the orientation of σ_H in scenario 2, this explains that more fractures show favorable orientations to open in that scenario.

However, the dilation potential of the reservoir fractures is not only controlled by the fracture strike but also their dip angle. In accordance to its definition (see Fig. 6.4), sub-vertical fractures tend to show highest dilation tendencies (e.g., Ferrill and Morris, 2003; Ferrill et al., 1999; Meixner et al., 2016; Moeck et al., 2009). Results of this study indicate that almost 97% (377 of 390) of those fractures with Td values higher than 0.8 are steeply dipping with 70-90° for scenario 1. In scenario 2, 87% (480 of 552) of those fractures are steeply dipping. In both scenarios, fractures with dip angles <45° can have maximum dilation tendencies of 0.6.

6.6.1.2 Fracture orientation and dip angle vs Ts

Regarding the slip tendency of the fractures, similar results were calculated indicating highest slip potential for orientations paralleling (+/-10°) σ_H, and lowest for orientations perpendicular to σ_H, respectively. Results of former studies would have led to an expectation of maximum values in directions 20-30° to σ_H (Barton et al., 1995; Rogers, 2003). Shear, however, will not occur on steep fracture planes parallel to the main stress directions, but can occur on conjugate sets of two planes, whose intersection angle is bisected by the main stress direction (here: σ_v) (cf. Jaeger et al., 2009; Morris et al., 1996; Secor, 1965). Therefore, results of both stress scenarios indicate lowest slip tendency values for fractures with sub-vertical to vertical orientation (see Figs. 6.6 and 6.7). As the slip tendency, as calculated in this study, is defined as the friction coefficient μ for cohesionless surfaces, potential sliding of a fracture depends on that coefficient. Byerlee (1978) stated that expected values for static friction generally range between 0.6 – 0.85, and thus, slip tendency values ≥ 0.6 indicate planes with ideal orientation for slip (Ferrill et al., 1999; Neves et al., 2009). For scenario 1, results indicate that only about 10 fractures show slip tendency values higher than 0.6, accounting for about 1% of the total reservoir fractures. If a slightly wider cut-off is chosen, defined by a normalized Ts' of 0.8, 64 fractures show slip tendencies in this range, which corresponds to 6% of the total amount of fractures. In contrast, using the stress regime of scenario 2, already almost 8% (85) of all fractures show slip tendencies higher than 0.6, and about one third (350) of all fractures have normalized slip tendencies higher than 0.8. Similar to the results of dilation tendency, variations in slip tendency can only depend on σ_2 as σ_H, as σ_1 is oriented vertical in both scenarios. In scenario 2, more fractures are favorably oriented relative to the maximum horizontal stress, which is evident by the mean strike orientation of all reservoir fractures being approximately NE – SW (see Figs. 6.5 and 6.7).

6.6.1.3 Derived transmissibility directions

Based on the hypothesis that fluid flow in the tight carbonate reservoir is mainly controlled by the fracture network, a general direction of transmissibility can be inferred for each stress scenario. For scenario 1, it has a main direction of 145° and preferentially fractures steeper than 60° would be reactivated. For scenario 2, the main direction is 40°, which is in good accordance with the mean strike orientation of 47° of the reservoir fractures (see Fig. 6.5). However, an anisotropic transmissivity with enhanced fluid flow in σ_H direction can be assumed for both scenarios, which fits the general assumptions stated by Ferrill et al. (1999).

In general, analyses indicate that, with regard to the defined stress scenarios, the transmissivity of the reservoir fractures is controlled by both, dip direction and dip angle, but with a greater influence of the dip angle. The difference within one scenario can therefore vary more widely for fractures with various dip angles than compared to two different stress scenarios with exemplarily only steeply dipping fractures. To prove this result, a comparison with production data would be essential.

6.6.2 Implications for reservoir modeling

Although documenting the various parameters that may influence how fractures could affect fluid flow is challenging, some key parameters can be calculated. With subsurface wireline-survey information from horizontal wells on fracture attitude and orientation, and in situ stress information it is possible to calculate the dilation tendency of fractures, which ultimately impacts flow properties in open fractures.

Apart from the fracture orientation and – density, which are essential input parameters for fractured reservoir models (e.g., Gillespie et al., 2011; Wennberg et al., 2016), the opening width or aperture of a fracture and the lateral extent control reservoir performance (e.g., Laubach, 2003; Nelson, 2001; Rogers, 2003). While the first parameters have been shown to be successfully incorporated from studies on suitable outcrop analogs (Becker et al., 2018; Bertotti et al., 2009; Lamarche et al., 2012; Wennberg et al., 2006), and thus, could be provided prior to actual well drilling for modeling the expected subsurface fracture network, the latter are controlled in part by in-situ stress conditions (Barton et al., 1995; Ferrill et al., 1999; Laubach, 2003; Rogers, 2003; Zeng et al., 2015). Hence, apertures derived from analogous surface rocks provide an upper limit of fracture characteristics as modeling input. This is due to a potential lateral extension of the rocks at the surface, where σ_h=0, which may cause wider fracture apertures (cf. Becker et al., 2018).

For infinite fractures, fracture permeability is proportional to the cube of the width (Brown, 1987; Lavrov, 2017; Marrett, 1997; Nelson, 2001). Here we present a simple model calculation neglecting actual fracture surface roughness: For a dilation tendency of 1, it would be assumed that the maximum possible width of a fracture is actually present at depth, resulting in the highest possible fracture permeability for that input. A dilation tendency of 1 results in 100% possible fracture permeability, i.e. the present day stress would not affect the flow along the fractures. However, this is rarely the case due to differently oriented fractures which would result in a strong overestimation of production performance of a reservoir. In the present study, calculations of scenario 1 result in an average dilation tendency of 0.65, indicating that in this case, the potential aperture width of the reservoir fractures would be reduced (Td = 0.65: 65% of the maximum possible fracture width is assumed to be conductive at depth). Calculating the actual fracture permeability with this reduced potential fracture width, this shows that only about 27% of the fracture permeability calculated for the defined maximum fracture width could be reached. In scenario 2, the average dilation tendency is 0.71, which would result in about 36% of the highest possible fracture permeability. Results of both scenarios show that neglecting the influence of the stress field would significantly overestimate production rates. Using the range of both results as a first approach for production performance estimation until actual flow rates are available limits the high range of uncertainties in conventional models.

Modeling the flow through the reservoir can be further optimized by integrating the position of fractures with calculated high dilation- or slip tendencies. This is exemplarily shown on an approximately 30 m long section of the studied well (Fig. 6.8). For scenario 1, 13 fractures show dilation tendencies above 0.8, which corresponds to a fracture intensity of 0.44 m^{-1} (Fig. 6.8a), and 11 fractures are critically stressed (Ts'>0.8), resulting in an intensity of 0.37 m^{-1}. For scenario 2, higher intensities of 0.74 m^{-1} correspond to the presence of 22 fractures with dilation tendencies above 0.8 (Fig. 6.8b), and likewise, a higher intensity of critically stressed (Ts'>0.8) fractures of 0.88 m^{-1} (26 fractures) occurs.

A further controlling factor is given by the shear potential of a fracture. Fracture surfaces are generally rough and not perfectly planar, with that roughness perfectly aligned forming the smallest possible opening width assuming an undisturbed state. If fractures are critically oriented for shearing, the rough surface will be slightly displaced and wider flow paths are formed (Rogers, 2003). Hence, the expected slip tendency of the fractures will add to the estimation of fracture permeability in the reservoir resulting in potentially higher values than if only the dilation tendency is used. Quantifying the amount of actual slip, and thus, resulting flow, as an appropriate rock failure criterion, e.g., depending on the rock strength, needs to be incorporated (cf. Barton et al., 1995; Rogers, 2003).

For the studied fractures, however, a flow system controlled by normal stress related processes, i.e. indicated by the dilation tendency, appears more likely due to maximum calculated results for Td and Ts are observed aligned to the σ_H orientation (cf. Rogers, 2003). Thus, for this setting, in a first approach to estimate fracture permeability only dilation tendency results are incorporated.

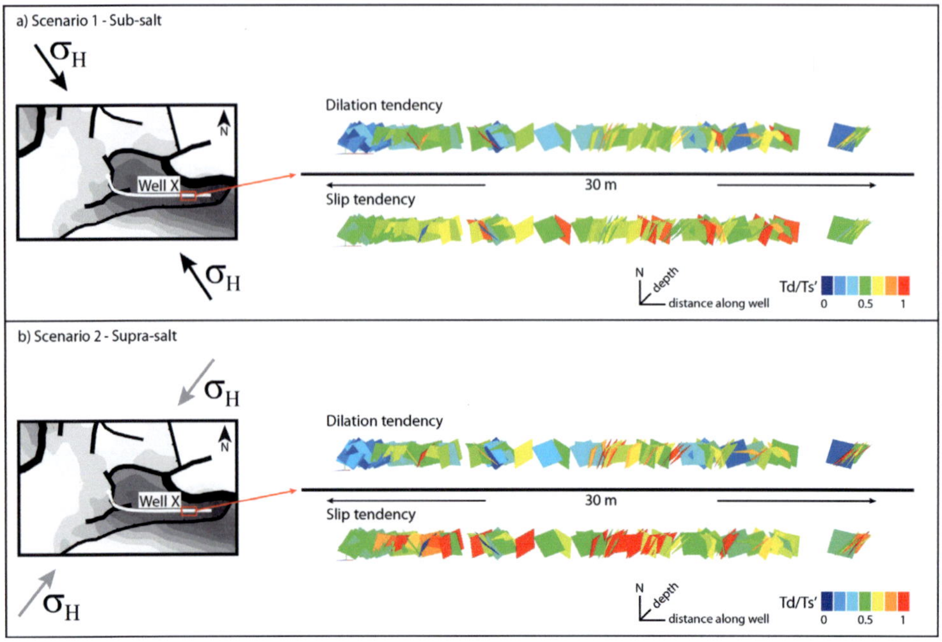

Figure 6.8. Fracture surfaces displayed along a defined 30 m long well section and color-coded according to their dilation- and slip tendencies as calculated for a) scenario 1, and b) scenario 2.

Diagenesis, i.e. minerals cementing the open fracture, may reduce fracture permeability even further, while partly sealed fractures may be kept open by cemented bridges even if unfavorably oriented to the present-day stress field (e.g., Gale et al., 2005; Gale et al., 2010; Hilgers et al., 2006; Hilgers et al., 2001; Laubach, 2003; Laubach et al., 2004; Olson et al., 2010; Prajapati et al., 2018). This effect incorporates different controlling factors as, exemplarily, the timing of fracture sealing events vs fracture reactivation, pore-pressure evolution, and rates of cement precipitation vs fracture opening rates (cf. Gale et al., 2005; Hilgers et al., 2006; Hilgers et al., 2001; Lander et al., 2002; Lander and Laubach, 2015; Laubach, 2003). As open and closed fractures can be well distinguished from the FMI-image log data, significant amounts of fracture-filling cements are not assumed for the studied open fractures. Although, without core material, no further constraints of potential cementation are possible in this study since uncertainties in the interpretation of FMI-image log data occur (e.g., Fernández-Ibáñez et al., 2018).

6.6.2.1 Uncertainty mitigation

The presented approach based on present stress data of nearby fields is valid for a first estimation of production performance. However, during the time of production, it should be noted that stress conditions are known to change (e.g., Morris and Ferrill, 2009; Segall and Fitzgerald, 1998; Teufel et al., 1991), and thus, estimations of the recovery potential should be accordingly adjusted.

With the available data, it cannot be stated whether scenario 1 or 2 are more likely to represent the actual reservoir conditions. In-situ stress measurements or production flow rates that could be correlated to the dilation tendencies would aid in gaining more accurate results and would support the aspired risk and uncertainty reduction. Due to the position of the reservoir at the base of the Zechstein cycles (see Fig. 6.3), below the Na2 and Na3 salt layers, that are reported to act as a mechanical decoupling horizon (Fleckenstein et al., 2004; Röckel and Lempp, 2003; Roth and Fleckenstein, 2001; Scheck and Bayer, 1999), scenario 1 seems more likely, though the abundance of open fractures aligned to the orientation of σ_H as in scenario 2 argues for the latter stress field (Fig. 6.8).

Furthermore, the studied field is located within a complex tectonic setting of repeatedly reactivated fault zones (see Fig. 6.2), partly initiated in pre-Permian times (e.g., Drozdzewski et al., 2009; Lohr et al., 2007). Becker et al. (2018) recently discussed the formation history of the studied fracture orientations in the context of such reactivated stress regimes and suggested local variations in mean fracture orientation to result from stress perturbations in the vicinity of faults that bound the studied reservoir (cf. Henk, 2005; Yale, 2003) (see Fig. 6.2b). Thus, the presently observed variations in the orientation of the maximum horizontal stress might also be related to those old structures explaining the local variations. Still, available published fracture and stress data cannot yet explain the regional tectonic context.

Results show that variations in Td and Ts are greater with respect to various dip angle than between the two different stress regimes. For the studied field in North Germany, where the magnitudes of σ_h and σ_H are close, this indicates that the true orientation of σ_H is less important for a first estimation of potential reservoir performance. In general, both scenarios indicate preferentially W – E oriented horizontal well orientations for NW Germany.

6.7 Conclusions

This study on fracture data of a horizontal well through Late Permian Zechstein Ca2 carbonates indicates the strong control of the present-day stress field on subsurface fracture characteristics. Based on knowledge of variations in the stress orientation in North Germany, two stress scenarios were defined for calculations of the dilation- and slip tendencies of those fractures.

Results of scenario 2 with NE – SW orientation of σ_H generally indicate better dilation tendencies than results of scenario 1 with a σ_H orientation of NW – SE. Variations in dilation- and slip tendency are larger with regard to the different dip angles of the fractures than due to the change in the orientation of the principal horizontal stress.

Transferring the calculated influence of the stress field on the opening potential of the studied reservoir fractures and its effect on the fracture permeability results in a reduced flow potential with 27% of the fractures potentially contributing to flow in scenario 1 and 36% in scenario 2. Thus, incorporation of stress data is key to estimate reservoir production performance.

7 Conclusions and Outlook

7.1 Conclusions

This study analyzes selected controlling factors on the reservoir quality of Upper Carboniferous fluvial sandstones and Zechstein Ca2 carbonates in NW Germany. For both reservoir settings, suitable outcrop analogs are defined and compared to well data of the associated reservoirs.

In Upper Carboniferous fluvial sandstones, the influence of diagenetic mineral formation and dissolution processes is linked to variations in matrix porosity and -permeability.

Two outcrop analogs at the southern margin of the Lower Saxony Basin, the lower Westphalian D Woitzel quarry and the upper Westphalian D Hüggel outcrop, indicate that reservoir characteristics depend on the stratigraphic position and sandstone deposition. The upper Westphalian D Hüggel sandstones show increased matrix permeabilities by one order of magnitude of average 0.54 mD compared to an average value of 0.03 mD for the lower Westphalian D Woitzel sandstones. This variation is related to the main type of authigenic clay mineral cement, which is represented by mainly kaolinite in the Hüggel sandstones and illite in the Woitzel sandstones. Humid depositional conditions during the early Westphalian D and associated higher original feldspar contents favored the processes of early kaolinitization followed by illitization in contrast to arid conditions during the late Westphalian D.

On a lateral scale of 150 m, permeabilities vary by up to two orders of magnitude within a single sandstone unit. On a tens of meters vertical scale, a positive correlation between average grain size and permeability occurs but is limited to samples with permeability values above 0.01 mD.

Results of those outcrop sandstones are compared to samples of two wells, Well A and B, of two gas fields located at a distance of approximately 50 km to the NE. Matrix permeabilities are shown to be similar to the outcrop values (0.01 to 10 mD) and generally controlled by authigenic clay cementation. However, matrix porosities strongly differ between the studied areas. The main controlling factor is the thermal exposure during burial resulting in different amounts of quartz cements. Lesser quartz cement development of Well B, as a response to Upper Cretaceous uplift, compared to Well A explains the better matrix porosities of the first. In addition, further porosity enhancement of the outcrop samples is related to the uplift to the surface and associated carbonate cement dissolution processes.

For the Upper Carboniferous sandstones in NW Germany, best reservoir characteristics are expected for areas strongly influenced by basin inversion related uplift processes ensuring a low thermal exposure and thus, lesser porosity reducing quartz cementation processes. Furthermore, open natural fractures can strongly enhance reservoir quality. Thermally overprinted areas or faults and erosion processes affecting the top seal integrity display major exploration risks.

With regard to the studied Zechstein Ca2 carbonate reservoirs of NW Germany, the focus is on the presence of open natural fractures enabling production of that formation by the development of a dual permeability system.

At first, a suitable outcrop analog for the studied horizontal well of a gas-producing Ca2 reservoir in the Lower Saxony Basin is defined. Three Zechstein Ca2 outcrops are therefore analyzed, which are located at a distance of approximately 130 km to the reservoir on the SW margin of the Harz mountains. The exposed carbonates are classified into three interpreted gross depositional environment types related to the slope, platform margin, and inner platform. Their lateral distribution is related to their position on the Eichsfeld-Altmark Swell.

Highest measured matrix porosities and –permeabilities are associated to platform margin carbonates. In contrast, slope carbonates have poorer reservoir qualities. Porosities appear to be comparable to subsurface slope samples, but permeabilities are increased by two orders of magnitude. With regard to the distribution on the paleo-high, reservoir quality is better in the central platform depositional types than in the more distal parts.

Based on the gross depositional environment classification, the Ührde quarry is chosen as a suitable analog for the studied Ca2 well, comprising a similar slope setting, a similar lithology and rock texture of dolo-mudstones to dolo-wackestones, and a comparable diagenetic evolution. Furthermore, uplift related fracturing is shown to be negligible and bitumen-impregnation is present on persistent fracture surfaces in the quarry. Hence, outcrop and subsurface fracture characteristics are comparable.

Fracture orientation and characteristics from the outcrop are measured by manual means and terrestrial laser scanning. The t-LiDAR workflow enables the generation of large amounts of data for inaccessible areas and is successfully validated by conventional compass measurements for accessible areas. Generally, dominant northerly striking fracture orientations are inferred from outcrop and subsurface fracture data. True mean orientations differ slightly which might be influenced by near-field stress perturbations around faults or the outcrop location on the NNE – SSW striking paleo-high. In addition, fracture intensities derived from t-LiDAR and well data are shown to be similar (t-LiDAR: 2.6 m^{-1}; well: 2.7 m^{-1}), but both appear biased in contrast to manually measured fracture data. For the t-LiDAR, fractures perpendicular to the wall orientation appear under-detected, while fractures paralleling the well are similarly underrepresented in the subsurface data set. Those results indicate a strong need for the incorporation of manual outcrop fracture data for data validation before using them as input for reservoir models.

A further controlling factor is represented by the present-day stress field influencing the opening potential of subsurface fractures, and thus, ultimately the flow pathways. For the studied horizontal well, two different stress scenarios are defined based on available stress data from northern Germany to derive the dilation- and slip tendency for the subsurface reservoir fractures.

The calculated results depend on the fracture dip angle and dip direction. For scenario 1, with a defined NW – SE orientation of the principal horizontal stress, lower dilation tendencies are determined compared to scenario 2 with a principal horizontal stress of NE – SW orientation. With regard to the flow potential, about 27% of the maximum possible fracture permeability would be expected for scenario 1, while 36% could be reached for scenario 2.

In conclusion, high reservoir quality intervals for both studied reservoir units, the Upper Carboniferous sandstones and the Ca2 carbonates, are related to naturally fractured areas associated to strong basin inversion tectonics. Pervasive cementation processes due to anomalous high temperature overprints reducing the matrix reservoir characteristics indicate a major risk especially in the Upper Carboniferous sandstones. The present-day stress field controls the fracture opening potential and thus, should be incorporated in reservoir modeling.

7.2 Outlook

For the Upper Carboniferous sandstones, the thermal exposure is shown to act as a main controlling factor for reservoir quality by influencing the amount of porosity reducing quartz cement formation. However, the calculated thermal indicator of this work (chapter 3) is based on a published burial history model for the Lower Saxony Basin (Bruns et al., 2013). Thus, more precise burial histories for the different studied areas might improve the results and would further enable to integrate the data into a reservoir quality prediction modeling software as Touchstone™.

This study illustrates that the presence of natural open fractures and their geometry controls reservoir quality, and thus, ultimately the productivity of the NW German Upper Carboniferous and Zechstein Ca2 reservoirs. The two case studies presented in this work indicate a regional trend of similar fracture orientations for northern Germany with local variations that might be related to the presence of large structural elements, i.e. faults.

However, for the Upper Carboniferous sandstones, fracture data is only available for the outcrops and should be analyzed and added for the subsurface to compare the fracture orientations. With regard to the Zechstein Ca2, this study only deals with the comparison of outcrop and subsurface fracture data of a carbonate slope environment (chapter 5). Similar to this approach, fracture data of the platform environment need to be analyzed. In addition, the regional fault patterns and associated processes, as exemplarily near-field stress perturbations (e.g., Henk, 2005), need further analysis. Linking the trend of such large-scale faults with fracture data from reservoirs and outcrops can assist prediction of variations in fracture orientations within one field, and thus, improve horizontal well planning. As especially the Upper Carboniferous bears great exploration potential (e.g. Besly, 1998; Jones and Glover, 2005; Kombrink et al., 2010; Leeder and Hardman, 1990), new information about the expected fluid migration pathways would significantly improve future development strategies.

Results of the flow potential of the fractures as shown in this work (chapter 6) can improve flow simulations within a reservoir, as fracture permeability varies depending on the orientation of the fractures with regard to the present-day stress field. No subsurface stress data has been available for the studied field and the calculated scenarios are based on stress data of a nearby field. Hence, incorporating stress data of the actual well might improve the results. Furthermore, subsurface stress data of the reservoir units of the Upper Carboniferous and Zechstein Ca2 should be generally added to a regional fault and fracture study to link potential local variations in fracture orientation to the stress field.

References

Albrecht, D., Reitenbach, V., 2015. Investigations on fluid transport properties in the North-German Rotliegend tight gas sandstones and applications. Environmental Earth Sciences 73, 5791-5799.

Amann-Hildenbrand, A., Dietrichs, J.P., Krooss, B.M., 2016. Effective gas permeability of tight gas sandstones as a function of capillary pressure - a non-steady-state approach. Geofluids 16, 367-383.

Baldschuhn, R., Binot, F., Fleig, S., Kockel, F., 2001. Geotektonischer Atlas von Nordwest-Deutschland und dem deutschen Nordsee-Sektor. Geologisches Jahrbuch (A) 153, 3-95.

Baldschuhn, R., Kockel, F., 1999. Das Osning-Lineament am Südrand des Niedersachsen-Beckens. Zeitschrift der Deutschen Geologischen Gesellschaft 150, 673-695.

Barton, C.A., Zoback, M.D., Moos, D., 1995. Fluid flow along potentially active faults in crystalline rock. Geology 23, 683-686.

Bastesen, E., Braathen, A., 2010. Extensional faults in fine grained carbonates – analysis of fault core lithology and thickness–displacement relationships. Journal of Structural Geology 32, 1609-1628.

Bastesen, E., Braathen, A., Skar, T., 2013. Comparison of scaling relationships of extensional fault cores in tight carbonate and porous sandstone reservoirs. Petroleum Geoscience 19, 385-398.

Beaufort, D., Cassagnabere, A., Petit, S., Lanson, B., Berger, G., Lacharpagne, J., Johansen, H., 1998. Kaolinite-to-dickite reaction in sandstone reservoirs. Clay minerals 33, 297- 316.

Becker, I., Koehrer, B., Waldvogel, M., Jelinek, W., Hilgers, C., 2018. Comparing fracture statistics from outcrop and reservoir data using conventional manual and t-LiDAR derived scanlines in Ca2 carbonates from the Southern Permian Basin, Germany. Marine and Petroleum Geology 95, 228-245.

Becker, I., Wüstefeld, P., Koehrer, B., Felder, M., Hilgers, C., 2017a. Porosity and permeability variations in a tight gas sandstone reservoir analogue, Westphalian D, Lower Saxony Basin, NW Germany: Influence of depositional setting and diagenesis. Journal of Petroleum Geology 40, 363-389.

Becker, I., Wüstefeld, P., Koehrer, B., Jelinek, W., Hilgers, C., 2017b. Digitally derived fracture network relationships in fractured Zechstein carbonate reservoirs from surface and well data. 79th EAGE Conference and Exhibition 2017.

Below, A., 1992. Fazies- und geochemische Diagenesestudie im Zechstein 2-Karbonat (Ca2) Nordwestdeutschlands: Facies- and geochemical diagenesis study of Zechstein 2-carbonates (Ca2) North-West-Germany, PhD thesis. Christian-Albrechts-Universität Kiel, Berichte-Reports Geol.-Paläont. Inst. Univ. Kiel, pp. 137.

Bertier, P., Swennen, R., Lagrou, D., Laenen, B.E.N., Kemps, R., 2008. Palaeo-climate controlled diagenesis of the Westphalian C & D fluvial sandstones in the Campine Basin (north-east Belgium). Sedimentology 55, 1375-1417.

Bertotti, G., Boro, H., Hardebol, N., Luthi, S.M., 2009. Improving predictions on mechanical stratigraphy of buried sedimentary successions: Lessons and workflow from outcrop analogs. AAPG Annual Conference and Exhibition, Denver, Colorado.

Besly, B., 1998. Carboniferous, in: Glennie, K. (Ed.), Petroleum geology of the North Sea: Basic concepts and recent advances. fourth ed. Blackwell ScienceLtd., Oxford, United Kingdom, fourth ed. Blackwell Science Ltd., Oxford, United Kingdom, pp. 104-136.

Besly, B., Burley, S., Turner, P., 1993. The late Carboniferous 'Barren Red Bed' play of the Silver Pit area, southern North Sea. Geological Society, London, Petroleum Geology Conference series, pp. 727-740.

Besly, B., Fielding, C., 1989. Palaeosols in Westphalian coal-bearing and red-bed sequences, central and northern England. Palaeogeography, Palaeoclimatology, Palaeoecology 70, 303-330.

Betz, D., Führer, F., Greiner, G., Plein, E., 1987. Evolution of the Lower Saxony Basin. Tectonophysics 137, 127-170.

Biehl, B.C., Reuning, L., Schoenherr, J., Lüders, V., Kukla, P.A., 2016. Impacts of hydrothermal dolomitization and thermochemical sulfate reduction on secondary porosity creation in deeply buried carbonates: A case study from the Lower Saxony Basin, northwest Germany. AAPG Bulletin 100, 597-621.

Binot, F., Gerling, P., Hiltmann, W., Kockel, F., Wehner, H., 1993. The petroleum system in the Lower Saxony Basin, in: Spencer, A.M. (Ed.), Generation, Accumulation and Production of Europe's Hydrocarbons III. Springer, Berlin Heidelberg, Special Publication of the European Association of Petroleum Geoscientists No. 3, pp. 121-139.

Bisdom, K., Gauthier, B.D.M., Bertotti, G., Hardebol, N.J., 2014. Calibrating discrete fracture-network models with a carbonate three-dimensional outcrop fracture network: Implications for naturally fractured reservoir modeling. AAPG Bulletin 98, 1351-1376.

Bjørlykke, K., 1998. Clay mineral diagenesis in sedimentary basins - a key to the prediction of rock properties. Examples from the North Sea Basin. Clay Minerals 33, 15-34.

Bjørlykke, K., Aagaard, P., 1992. Clay minerals in North Sea sandstones, in: Houseknecht, D.W., Pittman, E.D. (Eds.), Origin, diagenesis, and petrophysics of clay minerals in sandstones. SEPM Society for Sedimentary Geology, SEPM Special Publication, pp. 65- 80.

Bjørlykke, K., Egeberg, P., 1993. Quartz cementation in sedimentary basins. AAPG Bulletin 77, 1538-1548.

Bloch, S., 1991. Empirical prediction of porosity and permeability in sandstones (1). AAPG Bulletin 75, 1145-1160.

Boles, J.R., Franks, S.G., 1979. Clay diagenesis in Wilcox sandstones of southwest Texas: Implications of smectite diagenesis on sandstone cementation. Journal of Sedimentary Research 49, 55-70.

Bridge, J.S., 2006. Fluvial facies models: Recent developments, in: Posamentier, H., Walker, R. (Eds.), Facies models revisited. Society for Sedimentary Geology, SEPM Special Publication, pp. 85-170.

Brown, S.R., 1987. Fluid flow through rock joints: The effect of surface roughness. Journal of Geophysical Research: Solid Earth 92, 1337-1347.

Bruns, B., di Primio, R., Berner, U., Littke, R., 2013. Petroleum system evolution in the inverted Lower Saxony Basin, northwest Germany: A 3D basin modeling study. Geofluids 13, 246-271.

Busch, B., Hilgers, C., Gronen, L., Adelmann, D., 2017. Cementation and structural diagenesis of fluvio-aeolian Rotliegend sandstones, northern England. Journal of the Geological Society 174, 855-868.

Busch, B., Hilgers, C., Lander, R.H., Bonnell, L.M., Adelmann, D., 2018. Reservoir quality and burial model evaluation by kinetic quartz and illite cementation modeling: Case study of Rotliegendes, north Germany. AAPG Bulletin 102, 293-307.

Byerlee, J., 1978. Friction of rocks. Pure and Applied Geophysics 116, 615-626.

Caine, J.S., Evans, J.P., Forster, C.B., 1996. Fault zone architecture and permeability structure. Geology 24, 1025-1028.

Cameron, N., Ziegler, T., 1997. Probing the lower limits of a fairway: Further pre-Permian potential in the southern North Sea, in: Ziegler, K., Turner, P., Daines, S.R. (Eds.), Petroleum geology of the Southern North Sea: Future potential. Geological Society, London, Special Publications, pp. 123-141.

Cawood, A.J., Bond, C.E., Howell, J.A., Butler, R.W.H., Totake, Y., 2017. LiDAR, UAV or compass-clinometer? Accuracy, coverage and the effects on structural models. Journal of Structural Geology 98, 67-82.

Chan, M.A., Parry, W., Bowman, J., 2000. Diagenetic hematite and manganese oxides and fault-related fluid flow in Jurassic sandstones, southeastern Utah. AAPG Bulletin 84, 1281- 1310.

Childs, C., Manzocchi, T., Walsh, J.J., Bonson, C.G., Nicol, A., Schöpfer, M.P.J., 2009. A geometric model of fault zone and fault rock thickness variations. Journal of Structural Geology 31, 117-127.

Choquette, P.W., Pray, L.C., 1970. Geologic nomenclature and classification of porosity in sedimentary carbonates. AAPG Bulletin 54, 207-250.

Chuhan, F.A., Bjørlykke, K., Lowrey, C.J., 2001. Closed-system burial diagenesis in reservoir sandstones: Examples from the Garn Formation at Haltenbanken area, offshore mid-Norway. Journal of Sedimentary Research 71, 15-26.

Clark, D., 1980. The diagenesis of Zechstein carbonate sediments. Contributions to Sedimentology 9, 167-203.

Clark, D., 1986. The distribution of porosity in Zechstein carbonates, in: Brooks, J., Goff, J.C., van Hoorn, B. (Eds.), Habitat of Paleozoic gas in N.W. Europe. Geological Society, London, Special Publications, pp. 121-149.

Collettini, C., Trippetta, F., 2007. A slip tendency analysis to test mechanical and structural control on aftershock rupture planes. Earth and Planetary Science Letters 255, 402-413.

Cowan, G., 1993. Identification and significance of aeolian deposits within the dominantly fluvial Sherwood Sandstone Group of the East Irish Sea Basin UK, in: North, C.P., Prosser, D.J. (Eds.), Characterization of fluvial and aeolian reservoirs. Geological Society, London, Special Publications, pp. 231-245.

Dai, J., Ni, Y., Gong, D., Feng, Z., Liu, D., Peng, W., Han, W., 2017. Geochemical characteristics of gases from the largest tight sand gas field (Sulige) and shale gas field (Fuling) in China. Marine and Petroleum Geology 79, 426-438.

David, C., Wong, T.-F., Zhu, W., Zhang, J., 1994. Laboratory measurement of compaction-induced permeability change in porous rocks: Implications for the generation and maintenance of pore pressure excess in the crust. Pure and Applied Geophysics 143, 425- 456.

David, D.G.F., 1987. Sandkörper in fluviatilen Sandsteinen des Unteren Westfal D (Oberkarbon) am Piesberg bei Osnabrück. Facies 17, 51-57.

David, F., 1990. Sedimentologie und Beckenanalyse im Westfal C und D des nordwestdeutschen Oberkarbons. Deutsche Wissenschaftliche Gesellschaft für Erdöl, Erdgas und Kohle (DGMK), pp. 270.

Dershowitz, W.S., Herda, H.H., 1992. Interpretation of fracture spacing and intensity, in: Tillerson, J.R., Waversik, W.R. (Eds.), Rock mechanics. Balkema, Rotterdam, pp. 757- 766.

Dickinson, W.R., Suczek, C.A., 1979. Plate tectonics and sandstone compositions. AAPG Bulletin 63, 2164-2182.

Dreybrodt, W., 1990. The role of dissolution kinetics in the development of karst aquifers in limestone: A model simulation of karst evolution. The Journal of Geology 98, 639-655.

Drozdzewski, G., 1985. Tiefentektonik der Ibbenbürener Karbonscholle, in: Drozdzewski, G., Engel, H., Wolf, R., Wrede, V. (Eds.), Beiträge zur Tiefentektonik westdeutscher Steinkohlenlagerstätten. Geologisches Landesamt Nordrhein-Westfalen, pp. 189-216.

Drozdzewski, G., 1988. Die Wurzel der Osning-Überschiebung und der Mechanismus herzynischer Inversionsstörungen in Mitteleuropa. Geologische Rundschau 77, 127-141.

Drozdzewski, G., 2005. Zur sedimentären Entwicklung des Subvariscikums im Namurium und Westfalium Nordwestdeutschlands, in: Wrede, V. (Ed.), Stratigraphie von Deutschland V: Das Oberkarbon (Pennsylvanium) in Deutschland. pp. 271-320.

Drozdzewski, G., Henscheid, S., Hoth, P., Juch, D., Littke, R., Vieth, A., Wrede, V., 2009. The pre-Permian of NW-Germany – structure and coalification map. Zeitschrift der Deutschen Gesellschaft für Geowissenschaften 160, 159-172.

Dutton, S.P., Loucks, R.G., 2010. Diagenetic controls on evolution of porosity and permeability in lower Tertiary Wilcox sandstones from shallow to ultradeep (200–6700m) burial, Gulf of Mexico Basin, U.S.A. Marine and Petroleum Geology 27, 69-81.

Ehrenberg, S., 1991. Kaolinized, potassium-leached zones at the contacts of the Garn Formation, Haltenbanken, mid-Norwegian continental shelf. Marine and Petroleum Geology 8, 250-269.

Engelder, T., 1985. Loading paths to joint propagation during a tectonic cycle: An example from the Appalachian Plateau, USA. Journal of Structural Geology 7, 459-476.

English, J.M., 2012. Thermomechanical origin of regional fracture systems. AAPG Bulletin 96, 1597-1625.

English, J.M., Laubach, S.E., 2017. Opening-mode fracture systems: Insights from recent fluid inclusion microthermometry studies of crack-seal fracture cements, in: Turner, J.P., Healy, D., Hillis, R.R., Welch, M.J. (Eds.), Geomechanics and Geology. Geological Society, London, Special Publications, pp. 257-272.

Estupiñán, J.L., Marfil, R.P., Delgado, A.H., Permanyer, A., 2007. The impact of carbonate cements on the reservoir quality in the Napo Fm sandstones (Cretaceous Oriente Basin, Ecuador). Geologica Acta 5, 89-107.

Fall, A., Eichhubl, P., Bodnar, R.J., Laubach, S.E., Davis, J.S., 2015. Natural hydraulic fracturing of tight-gas sandstone reservoirs, Piceance Basin, Colorado. Geological Society of America Bulletin 127, 61-75.

Fernández-Ibáñez, F., DeGraff, J., Ibrayev, F., 2018. Integrating borehole image logs with core: A method to enhance subsurface fracture characterization. AAPG Bulletin 102, 1067- 1090.

Ferrill, D.A., Morris, A.P., 2003. Dilational normal faults. Journal of Structural Geology 25, 183-196.

Ferrill, D.A., Winterle, J., Wittmeyer, G., Sims, D., Colton, S., Armstrong, A., Morris, A.P., 1999. Stressed rock strains groundwater at Yucca Mountain, Nevada. GSA Today 9, 1-8.

Fleckenstein, P., Reuschke, G., Müller, B., Connolly, P., 2004. Predicting stress re-orientations associated with major geological structures in sedimentary sequences, Tight gas reservoirs - natural gas for the future, DGMK-Research Report 593-5, pp. 95.

Flügel, E., 2010. Microfacies of carbonate rocks, Springer-Verlag Berlin Heidelberg, pp. 984.

Franzke, H., Müller, R., Voigt, T., von Eynatten, H., 2007. Paleo-stress paths in the Harz Mountains and surrounding areas (Germany) between the Triassic and the Upper Cretaceous. Zeitschrift für Geologische Wissenschaften 35, 141-156.

Füchtbauer, H., 1974. Zur Diagenese fluviatiler Sandsteine. Geologische Rundschau 63, 904-925.

Füchtbauer, H., 1988. Sandsteine, in: Füchtbauer, H. (Ed.), Sedimente und Sedimentgesteine. Schweizerbart, Stuttgart, Germany, Schweizerbart, pp. 94-187.

Füchtbauer, H., 1998. Sedimente und Sedimentgesteine, Schweizerbart, Stuttgart, pp. 1141.

Gaich, A., Poetsch, M., Schubert, W., 2006. Acquisition and assessment of geometric rock mass features by true 3D images. Golden Rocks 2006, The 41st US Symposium on Rock Mechanics (USRMS). American Rock Mechanics Association.

Gale, J.F., Laubach, S.E., Olson, J.E., Marrett, R., 2005. Using the link between diagenesis and fracturing to accurately predict, characterize, and model fluid-flow in fractured carbonate rocks. SPE Latin American and Caribbean Petroleum Engineering Conference. Society of Petroleum Engineers.

Gale, J.F.W., Lander, R.H., Reed, R.M., Laubach, S.E., 2010. Modeling fracture porosity evolution in dolostone. Journal of Structural Geology 32, 1201-1211.

García-Sellés, D., Falivene, O., Arbués, P., Gratacos, O., Tavani, S., Muñoz, J.A., 2011. Supervised identification and reconstruction of near-planar geological surfaces from terrestrial laser scanning. Computers & Geosciences 37, 1584-1594.

Garland, J., Neilson, J.E., Laubach, S.E., Whidden, K.J., 2012. Advances in carbonate exploration and reservoir analysis, in: Garland, J., Neilson, J.E., Laubach, S.E., Whidden, K.J. (Eds.), Advances in carbonate exploration and reservoir analysis. Geological Society, London, Special Publications, pp. 1-15.

Garzanti, E., Vezzoli, G., 2003. A classification of metamorphic grains in sands based on their composition and grade. Journal of Sedimentary Research 73, 830-837.

Gast, R., Gundlach, T., 2006. Permian strike slip and extensional tectonics in Lower Saxony, Germany. Zeitschrift der Deutschen Gesellschaft für Geowissenschaften 157, 41- 55.

Gaupp, R., Matter, A., Platt, J., Ramseyer, K., Walzebuck, J., 1993. Diagenesis and fluid evolution of deeply buried Permian (Rotliegende) gas reservoirs, northwest Germany. AAPG Bulletin 77, 1111-1128.

Gaupp, R., Okkerman, J.A., 2011. Diagenesis and reservoir quality of Rotliegend sandstones in the northern Netherlands - a review, in: Grötsch, J., Gaupp, R. (Eds.), The Permian Rotliegend of the Netherlands. SEPM Society for Sedimentary Geology, pp. 193- 226.

Geluk, M., 2000. Late Permian (Zechstein) carbonate-facies maps, the Netherlands. Netherlands Journal of Geosciences 79, 17-27.

Gigli, G., Casagli, N., 2011. Semi-automatic extraction of rock mass structural data from high resolution LIDAR point clouds. International Journal of Rock Mechanics and Mining Sciences 48, 187-198.

Gillespie, P., Monsen, E., Maerten, L., Hunt, D., Thurmond, J., Tuck, D., 2011. Fractures in carbonates: From digital outcrops to mechanical models. Outcrops revitalized—Tools, techniques and applications: Tulsa, Oklahoma, SEPM Concepts in Sedimentology and Paleontology 10, 137-147.

Glennie, K., 1986. Development of NW Europe's Southern Permian gas basin, in: Brooks, J., Goff, J.C., van Hoorn, B. (Eds.), Habitat of Paleozoic Gas in N.W. Europe. Geological Society, London, Special Publications, pp. 3-22.

Glennie, K., Buller, A., 1983. The Permian Weissliegend of NW Europe: The partial deformation of aeolian dune sands caused by the Zechstein transgression. Sedimentary Geology 35, 43-81.

Glennie, K., Underhill, J., 1998. Origin, development and evolution of structural styles, in: Glennie, K. (Ed.), Petroleum Geology of the North Sea: Basic Concepts and Recent Advances. Oxford, United Kingdom, Blackwell Science, Oxford, United Kingdom, Blackwell Science, pp. 42-84.

Grünthal, G., 2004. Erdbeben und Erdbebengefährdung in Deutschland sowie im europäischen Kontext. Geographie und Schule 151, 14-23.

Grünthal, G., Mayer-Rosa, D., Lenhardt, W., 1998. Abschätzung der Erdbebengefährdung für die D-A-CH-Staaten - Deutschland, Österreich, Schweiz. Bautechnik 75, 753-767.

Haglauer-Ruppel, B., 1991. Kinematik und Begleitstrukturen von Scherzonen: Experimente und Beispiele Mitteleuropas (mit besonderer Berücksichtigung des Osning-Lineamentes), PhD thesis. Ruhr University Bochum, pp. 223.

Hancock, N., Taylor, A., 1978. Clay mineral diagenesis and oil migration in the Middle Jurassic Brent Sand Formation. Journal of the Geological Society 135, 69-72.

Haughton, P., Todd, S., Morton, A., 1991. Sedimentary provenance studies, in: Morton, A., Todd, S., Haughton, P. (Eds.), Developments in sedimentary provenance studies. Geological Society, London, Special Publications, Geological Society, London, Special Publications, pp. 1-11.

Henk, A., 2005. Pre-drilling prediction of the tectonic stress field with geomechanical models. First Break 23, 53-57.

Hennings, P., Allwardt, P., Paul, P., Zahm, C., Reid, R., Alley, H., Kirschner, R., Lee, B., Hough, E., 2012. Relationship between fractures, fault zones, stress, and reservoir productivity in the Suban gas field, Sumatra, Indonesia. AAPG Bulletin 96, 753-772.

Herrmann, A., 1956. Der Zechstein am südwestlichen Harzrand. Geol. Jb 72, 1-72.

Higgs, K.E., Zwingmann, H., Reyes, A.G., Funnell, R.H., 2007. Diagenesis, porosity evolution, and petroleum emplacement in tight gas reservoirs, Taranaki Basin, New Zealand. Journal of Sedimentary Research 77, 1003-1025.

Hilgers, C., Kirschner, D.L., Breton, J.P., Urai, J.L., 2006. Fracture sealing and fluid overpressures in limestones of the Jabal Akhdar dome, Oman mountains. Geofluids 6, 168 - 184.

Hilgers, C., Koehn, D., Bons, P., Urai, J., 2001. Development of crystal morphology during unitaxial growth in a progressively widening vein: II. Numerical simulations of the evolution of antitaxial fibrous veins. Journal of Structural Geology 23, 873-885.

Hodgetts, D., 2013. Laser scanning and digital outcrop geology in the petroleum industry: A review. Marine and Petroleum Geology 46, 335-354.

Holditch, S.A., 2006. Tight gas sands. Journal of Petroleum Technology 58, 86-93.

Hollmann, G., Klug, B., Schmitz, J., Stahl, E., Wellens, M., 1997. Schneeren-Husum - zur Geologie einer Erdgaslagerstätte im Nordwestdeutschen Oberkarbon. Niedersächs. Akad. der Geowiss. 13, 33-43.

Holloway, S., Rochelle, C., Bateman, K., Pearce, J., Baily, H., Metcalfe, R., 1996. The underground disposal of carbon dioxide. British Geological Survey, Final report of Joule II project no. CT92-0031, pp. 323.

Hood, K., Yurewicz, D., 2008. Assessing the Mesaverde basin-centered gas play, Piceance Basin, Colorado, in: Cumella, S., Shanley, K., Camp, W. (Eds.), Understanding, exploring, and developing tight-gas sands. 2005 Vail Hedberg Conference, AAPG Hedberg Series, pp. 87-104.

House, N., Shemeta, J., 2008. Understanding hydraulic fractures in tight-gas sands through the integration of borehole microseismic data, three-dimensional surface seismic data, and three-dimensional vertical seismic profile data: A Jonah field case study, in: Cumella, S., Shanley, K., Camp, W. (Eds.), Understanding, exploring, and developing tight-gas sands. AAPG Hedberg Series, 2005 Vail Hedberg Conference, pp. 77-86.

Hurst, A., Nadeau, P., 1994. Estimation of water saturation from clay microporosity data. European Petroleum Conference, Society of Petroleum Engineers, SPE 28850, pp. 451- 458.

Huttel, P., 1989. Das Stassfurt-Karbonat (Ca2) in Süd-Oldenburg: Fazies und Diagenese eines Sediments am Nordhang der Hunte-Schwelle. Geolog. Inst. d. Georg-August-Univ., pp. 94.

Jaeger, J.C., Cook, N.G., Zimmerman, R., 2009. Fundamentals of rock mechanics. John Wiley & Sons, Blackwell Publishing, pp. 469.

Jiang, Z.-X., Li, Z., Li, F., Pang, X.-Q., Yang, W., Liu, L.-F., Jiang, F.-J., 2015. Tight sandstone gas accumulation mechanism and development models. Petroleum Science 12, 587-605.

Jones, N.S., Glover, B.W., 2005. Fluvial sandbody architecture, cyclicity and sequence stratigraphical setting - implications for hydrocarbon reservoirs: The Westphalian C and D of the Osnabrück-Ibbenbüren area, northwest Germany, in: Collinson, J.D., Evans, D.J., Holliday, D.W., and Jones, N.S. (Ed.), Carboniferous hydrocarbon geology, the southern North Sea and surrounding onshore areas. Yorkshire Geological Society, Yorkshire Geological Society, pp. 57-74.

Jones, R.R., Kokkalas, S., McCaffrey, K.J.W., 2009a. Quantitative analysis and visualization of nonplanar fault surfaces using terrestrial laser scanning (LIDAR) - The Arkitsa fault, central Greece, as a case study. Geosphere 5, 465-482.

Jones, R.R., McCaffrey, K.J.W., Clegg, P., Wilson, R.W., Holliman, N.S., Holdsworth, R.E., Imber, J., Waggott, S., 2009b. Integration of regional to outcrop digital data: 3D visualisation of multi-scale geological models. Computers & Geosciences 35, 4-18.

Karnin, W.-D., Gast, R., Bärle, C., Clever, B., Kühn, M., Sommer, J., 2006. Play types, structural history and distribution of Middle Buntsandstein gas fields in NW Germany:

Observations and their genetic interpretation. Zeitschrift der Deutschen Gesellschaft für Geowissenschaften 157, 121-133.

Karnin, W.-D., Rockenbauch, K., Ruijtenberg, P., 1992. The effect of the succes of 3D seismic data on the exploration and appraisal of Zechstein targets in NW Germany. First Break 10, 233-240.

Karnin, W., Idiz, E., Merkel, D., Ruprecht, E., 1996. The Zechstein Stassfurt Carbonate hydrocarbon system of the Thuringian Basin, Germany. Petroleum Geoscience 2, 53-58.

Kemeny, J., Turner, K., Norton, B., 2006. LIDAR for rock mass characterization: hardware, software, accuracy and best-practices. in: Tonon, F., Kottenstette, J. (Eds.), Laser and photogrammetric methods for rock face characterization. ARMA, American Rock Mechanics Association, Golden, Colorado, pp. 49-62.

Ketzer, J., Morad, S., Amorosi, A., 2003. Predictive diagenetic clay-mineral distribution in siliciclastic rocks within a sequence stratigraphic framework, in: Worden, R.H., Morad, S. (Eds.), Clay mineral cements in sandstones. International Association of Sedimentologists Special Publication, pp. 43-61.

Khalifa, M.A., Morad, S., 2015. Impact of depositional facies on the distribution of diagenetic alterations in the Devonian shoreface sandstone reservoirs, Southern Ghadamis Basin, Libya. Sedimentary Geology 329, 62-80.

Kim, J.C., Lee, Y.I., Hisada, K.-i., 2007. Depositional and compositional controls on sandstone diagenesis, the Tetori Group (Middle Jurassic–Early Cretaceous), central Japan. Sedimentary Geology 195, 183-202.

Klassen, H., 1984. Geologie des Osnabrücker Berglandes. Naturwissenschaftliches Museum Osnabrück, pp. 672.

Kley, J., Voigt, T., 2008. Late Cretaceous intraplate thrusting in central Europe: Effect of Africa-Iberia-Europe convergence, not Alpine collision. Geology 36, 839.

Knott, S.D., Beach, A., Brockbank, P.J., Brown, J.L., McCallum, J.E., Welbon, A.I., 1996. Spatial and mechanical controls on normal fault populations. Journal of Structural Geology 18, 359-372.

Kockel, F., 2003. Inversion structures in Central Europe - Expressions and reasons, an open discussion. Netherlands Journal of Geosciences 82, 367-382.

Koehrer, B., Salchenegger, S., Degen, D., Althoff, J., Dreier, J., 2016. Modeling of a naturally fractured carbonate reservoir based on petrophysical rock types, EAGE 78th Conference and Exhibition 2016, Vienna, Austria.

Kombrink, H., 2008. The Carboniferous of the Netherlands and surrounding areas; a basin analysis. Geologica Ultraiectina (294), PhD thesis. Departement Aardwetenschappen, Universiteit Utrech, pp. 184.

Kombrink, H., Besley, B., Collinson, J., Den Hartog Jager, D., Drozdzewski, G., Dusar, M., Hoth, P., Pagnier, H., Stemmerik, L., Waksmundzka, M., Wrede, V., 2010. Carboniferous, in: Doornenbal, J.C., Stevenson, A.G. (Eds.), Petroleum geological atlas of the Southern Permian Basin area. EAGE Publications bv, Houten, pp. 81-99.

Kombrink, H., Bridge, J., Stouthamer, E., 2007. The alluvial architecture of the Coevorden Field (Upper Carboniferous), the Netherlands. Netherlands Journal of Geosciences 86, 3- 14.

Krawczyk, W.E., Ford, D.C., 2006. Correlating specific conductivity with total hardness in limestone and dolomite karst waters. Earth Surface Processes and Landforms 31, 221- 234.

Lamarche, J., Lavenu, A.P.C., Gauthier, B.D.M., Guglielmi, Y., Jayet, O., 2012. Relationships between fracture patterns, geodynamics and mechanical stratigraphy in carbonates (South-East Basin, France). Tectonophysics 581, 231-245.

Lander, R., Gale, J., Laubach, S., Bonnell, L., 2002. Interaction between quartz cementation and fracturing in sandstones. AAPG Annual Convention Program.

Lander, R.H., Larese, R.E., Bonnell, L.M., 2008. Toward more accurate quartz cement models: The importance of euhedral versus noneuhedral growth rates. AAPG Bulletin 92, 1537-1563.

Lander, R.H., Laubach, S.E., 2015. Insights into rates of fracture growth and sealing from a model for quartz cementation in fractured sandstones. Geological Society of America Bulletin 127, 516-538.

Lanson, B., Beaufort, D., Berger, G., Bauer, A., Cassagnabère, A., Meunier, A., 2002. Authigenic kaolin and illitic minerals during burial diagenesis of sandstones: A review. Clay Minerals 37, 1-22.

Laubach, S.E., 2003. Practical approaches to identifying sealed and open fractures. AAPG Bulletin 87, 561-579.

Laubach, S.E., Fall, A., Copley, L.K., Marrett, R., Wilkins, S.J., 2016. Fracture porosity creation and persistence in a basement-involved Laramide fold, Upper Cretaceous Frontier Formation, Green River Basin, USA. Geological Magazine 153, 887-910.

Laubach, S.E., Gale, J., 2006. Obtaining fracture information for low-permeability (tight) gas sandstones from sidewall cores. Journal of Petroleum Geology 29, 147-158.

Laubach, S.E., Olson, J.E., Gale, J.F.W., 2004. Are open fractures necessarily aligned with maximum horizontal stress? Earth and Planetary Science Letters 222, 191-195.

Laubach, S.E., Olson, J.E., Gross, M.R., 2009. Mechanical and fracture stratigraphy. AAPG Bulletin 93, 1413-1426.

Laux, D., Henk, A., 2015. Terrestrial laser scanning and fracture network characterisation – perspectives for a (semi-) automatic analysis of point cloud data from outcrops. Zeitschrift der Deutschen Gesellschaft für Geowissenschaften 166, 99-118.

Lavenu, A.P.C., Lamarche, J., Gallois, A., Gauthier, B.D.M., 2013. Tectonic versus diagenetic origin of fractures in a naturally fractured carbonate reservoir analog (Nerthe anticline, southeastern France). AAPG Bulletin 97, 2207-2232.

Lavrov, A., 2017. Fracture permeability under normal stress: A fully computational approach. Journal of Petroleum Exploration and Production Technology 7, 181-194.

LBEG, 2016. Erdöl und Erdgas in der Bundesrepublik Deutschland. Landesamt für Bergbau, Energie und Geologie, Hannover, Germany.

Leeder, M.R., Hardman, M., 1990. Carboniferous geology of the Southern North Sea Basin and controls on hydrocarbon prospectivity, in: Hardman, R.F.P., Brooks, J. (Eds.), Tectonic events responsible for Britain's oil and gas reserves. Geological Society, London, Special Publications, pp. 87-105.

Lisle, R.J., Srivastava, D.C., 2004. Test of the frictional reactivation theory for faults and validity of fault-slip analysis. Geology 32, 569.

Liu, Z., Yuan, D., Dreybrodt, W., 2005. Comparative study of dissolution rate-determining mechanisms of limestone and dolomite. Environmental Geology 49, 274-279.

Lohr, T., Krawczyk, C.M., Tanner, D.C., Samiee, R., Endres, H., Oncken, O., Trappe, H., Kukla, P.A., 2007. Strain partitioning due to salt: Insights from interpretation of a 3D seismic data set in the NW German Basin. Basin Research 19, 579-597.

Long, J.C.S., Witherspoon, P.A., 1985. The relationship of the degree of interconnection to permeability in fracture networks. Journal of Geophysical Research 90, 3087-3098.

Lucia, F.J., 2007. Carbonate reservoir characterization: An integrated approach, 2nd ed. Springer-Verlag Berlin Heidelberg, pp. 336.

Lüders, V., Plessen, B., di Primio, R., 2012. Stable carbon isotopic ratios of CH4–CO2-bearing fluid inclusions in fracture-fill mineralization from the Lower Saxony Basin (Germany) – A tool for tracing gas sources and maturity. Marine and Petroleum Geology 30, 174-183.

Lüders, V., Plessen, B., Romer, R.L., Weise, S.M., Banks, D.A., Hoth, P., Dulski, P., Schettler, G., 2010. Chemistry and isotopic composition of Rotliegend and Upper Carboniferous formation waters from the North German Basin. Chemical Geology 276, 198-208.

Lyu, W., Zeng, L., Zhang, B., Miao, F., Lyu, P., Dong, S., 2017. Influence of natural fractures on gas accumulation in the Upper Triassic tight gas sandstones in the northwestern Sichuan Basin, China. Marine and Petroleum Geology 83, 60-72.

Makowitz, A., Lander, R., Milliken, K., 2006. Diagenetic modeling to assess the relative timing of quartz cementation and brittle grain processes during compaction. AAPG Bulletin 90, 873-885.

Mandl, G., 1999. Faulting in brittle rocks: An introduction to the mechanics of tectonic faults. Springer Verlag Berlin Heidelberg, pp. 434.

Marrett, R., 1997. Permeability, porosity, and shear-wave anisotropy from scaling of open fracture populations, in: Hoak, T.E., Klawitter, A., Blomquist, P.K. (Eds.), Fractured reservoirs: Characterization and modeling. Rocky Mountain Association of Geologists, pp. 217-226.

McBride, E.F., 1963. A classification of common sandstones. SEPM Journal of Sedimentary Research 33, 664-669.

McKinley, J.M., Atkinson, P.M., Lloyd, C.D., Ruffell, A.H., Worden, R.H., 2011. How porosity and permeability vary spatially with grain size, sorting, cement volume, and mineral dissolution in fluvial Triassic sandstones: the value of geostatistics and local regression. Journal of Sedimentary Research 81, 844-858.

Meadows, N.S., Beach, A., 1993. Controls on reservoir quality in the Triassic Sherwood Sandstone of the Irish Sea. Geological Society, London, Petroleum Geology Conference series 4, 823-833.

Meixner, J., Schill, E., Grimmer, J.C., Gaucher, E., Kohl, T., Klingler, P., 2016. Structural control of geothermal reservoirs in extensional tectonic settings: An example from the Upper Rhine Graben. Journal of Structural Geology 82, 1-15.

Menning, M., Alekseev, A.S., Chuvashov, B.I., Davydov, V.I., Devuyst, F.X., Forke, H.C., Grunt, T.A., Hance, L., Heckel, P.H., Izokh, N.G., Jin, Y.G., Jones, P.J., Kotlyar, G.V., Kozur, H.W., Nemyrovska, T.I., Schneider, J.W., Wang, X.D., Weddige, K., Weyer, D., Work, D.M., 2006. Global time scale and regional stratigraphic reference scales of Central and West Europe, East Europe, Tethys, South China, and North America as used in the Devonian–Carboniferous–Permian Correlation Chart 2003 (DCP 2003). Palaeogeography, Palaeoclimatology, Palaeoecology 240, 318-372.

Miall, A.D., 1988. Architectural elements and bounding surfaces in fluvial deposits: anatomy of the Kayenta Formation (Lower Jurassic), southwest Colorado. Sedimentary Geology 55, 233-262.

Moeck, I., Kwiatek, G., Zimmermann, G., 2009. Slip tendency analysis, fault reactivation potential and induced seismicity in a deep geothermal reservoir. Journal of Structural Geology 31, 1174-1182.

Molenaar, N., Felder, M., Bär, K., Götz, A.E., 2015. What classic greywacke (litharenite) can reveal about feldspar diagenesis: An example from Permian Rotliegend sandstone in Hessen, Germany. Sedimentary Geology 326, 79-93.

Morad, S., Al-Ramadan, K., Ketzer, J.M., De Ros, L.F., 2010. The impact of diagenesis on the heterogeneity of sandstone reservoirs: A review of the role of depositional facies and sequence stratigraphy. AAPG Bulletin 94, 1267-1309.

Morad, S., Ketzer, J., De Ros, L.F., 2000. Spatial and temporal distribution of diagenetic alterations in siliciclastic rocks: implications for mass transfer in sedimentary basins. Sedimentology 47, 95-120.

Morris, A., Ferrill, D.A., Brent Henderson, D.B., 1996. Slip-tendency analysis and fault reactivation. Geology 24, 275-278.

Morris, A.P., Ferrill, D.A., 2009. The importance of the effective intermediate principal stress (σ'2) to fault slip patterns. Journal of Structural Geology 31, 950-959.

Morris, K., Shepperd, C., 1982. The role of clay minerals in influencing porosity and permeability characteristics in the Bridport Sands of Wytch Farm, Dorset. Clay Minerals 17, 41-54.

Moscariello, A., 2003. The Schooner Field, Blocks 44/26a, 43/30a, UK North Sea. Geological Society, London, Memoirs 20, 811-824.

Narr, W., 1991. Fracture density in the deep subsurface: Techniques with application to Point Arguello oil field. AAPG Bulletin 75, 1300-1323.

Narr, W., 1996. Estimating average fracture spacing in subsurface rock. AAPG Bulletin 80, 1565-1585.

Navabpour, P., Malz, A., Kley, J., Siegburg, M., Kasch, N., Ustaszewski, K., 2017. Intraplate brittle deformation and states of paleostress constrained by fault kinematics in the central German platform. Tectonophysics 694, 146-163.

Nelson, R.A., 2001. Geologic analysis of naturally fractured reservoirs. Gulf Professional Publishing, pp. 332.

Neves, M.C., Paiva, L.T., Luis, J., 2009. Software for slip-tendency analysis in 3D: A plug-in for Coulomb. Computers & Geosciences 35, 2345-2352.

Nolen-Hoeksema, R.C., Howard, J., 1987. Estimating drilling direction for optimum production in a fractured reservoir. AAPG Bulletin 71, 958-966.

Olivarius, M., Weibel, R., Hjuler, M.L., Kristensen, L., Mathiesen, A., Nielsen, L.H., Kjøller, C., 2015. Diagenetic effects on porosity–permeability relationships in red beds of the Lower Triassic Bunter Sandstone Formation in the North German Basin. Sedimentary Geology 321, 139-153.

Olson, J.E., Laubach, S.E., Eichhubl, P., 2010. Estimating natural fracture producibility in tight gas sandstones: Coupling diagenesis with geochemical modeling. The Leading Edge 29, 1494-1499.

Olson, J.E., Laubach, S.E., Lander, R.H., 2009. Natural fracture characterization in tight gas sandstones: Integrating mechanics and diagenesis. AAPG Bulletin 93, 1535-1549.

Ortega, O.J., Marrett, R.A., Laubach, S.E., 2006. A scale-independent approach to fracture intensity and average spacing measurement. AAPG Bulletin 90, 193-208.

Paul, J., 1986. Stratigraphy of the Lower Werra cycle (Z1) in West Germany (preliminary results), in: Harwood, G.M., Smith, D.B. (Eds.), The English Zechstein and related topics. Geological Society, London, Special Publications, pp. 149-156.

Paul, J., 1991. Zechstein carbonates-marine episodes of a hyperhaline sea. Zentralblatt für Geologie und Paläontologie Teil 1, 1029-1045.

Paul, J., 1993. Anatomie und Entwicklung eines permo-triassischen Hochgebietes: Die Eichsfeld-Altmark-Schwelle. Geologisches Jahrbuch (A: Allgemeine und regionale Geologie BR Deutschland und Nachbargebiete, Tektonik, Stratigraphie, Paläontologie) 131, 197-218.

Paul, J., 2006. The Kupferschiefer: Lithology, stratigraphy, facies and metallogeny of a black-shale. Zeitschrift der Deutschen Gesellschaft für Geowissenschaften, 57-76.

Paul, J., 2014. Gypsum domes and diapirs: Common features in the Zechstein (Upper Permian) of Germany. Geological Quarterly 58 (3), 521-530.

Paxton, S., Szabo, J., Ajdukiewicz, J., Klimentidis, R., 2002. Construction of an intergranular volume compaction curve for evaluating and predicting compaction and porosity loss in rigid-grain sandstone reservoirs. AAPG Bulletin 86, 2047-2067.

Peacock, D., Harris, S., Mauldon, M., 2003. Use of curved scanlines and boreholes to predict fracture frequencies. Journal of Structural Geology 25, 109-119.

Pearce, T., McLean, D., Wray, D., Wright, D., Jeans, C., Mearns, E., 2005. Stratigraphy of the Upper Carboniferous Schooner Formation, southern North Sea: chemostratigraphy, mineralogy, palynology and Sm-Nd isotope analysis, in: Collinson, J., Evans, D., Holliday, D., Jones, N. (Eds.), Carboniferous hydrocarbon geology: The southern North Sea and surrounding onshore areas. Yorkshire Geological Society, Occasional Publications series, pp. 165-182.

Perras, M.A., Diederichs, M.S., 2014. A review of the tensile strength of rock: Concepts and testing. Geotechnical and Geological Engineering 32, 525-546.

Peryt, T., Geluk, M., Mathiesen, A., Paul, J., Smith, K., 2010. Zechstein, in: Doornenbal, J.C., Stevenson, A.G. (Eds.), Petroleum geological atlas of the Southern Permian Basin area. EAGE Publications bv, Houten, pp. 123-147.

Petmecky, S., Meier, L., Reiser, H., Littke, R., 1999. High thermal maturity in the Lower Saxony Basin: intrusion or deep burial? Tectonophysics 304, 317-344.

Pharaoh, T., Dusar, M., Geluk, M., Kockel, F., Krawczyk, C., Krzywiec, P., Scheck-Wenderoth, M., Thybo, H., Vejbaek, O., van Wees, J.D., 2010. Tectonic evolution, in: Doornenbal, J.C., Stevenson, A.G. (Eds.), Petroleum geological atlas of the Southern Permian Basin area. EAGE Publications bv, Houten, pp. 25-57.

Philip, Z.G., Jennings, J.W.J., Olson, J.E., Laubach, S.E., Holder, J., 2005. Modeling coupled fracture-matrix fluid flow on geomechanically simulated fracture networks. SPE Reservoir Evaluation & Engineering 8, 300-309.

Prajapati, N., Selzer, M., Nestler, B., Busch, B., Hilgers, C., 2018. Modeling fracture cementation processes in calcite limestone: A phase-field study. Geothermal Energy 6.

Pranter, M.J., Ellison, A.I., Cole, R.D., Patterson, P.E., 2007. Analysis and modeling of intermediate-scale reservoir heterogeneity based on a fluvial point-bar outcrop analog, Williams Fork Formation, Piceance Basin, Colorado. AAPG Bulletin 91, 1025-1051.

Price, N.J., Cosgrove, J.W., 1990. Analysis of geological structures. Cambridge University Press, pp. 520.

Pusch, G., Gaupp, R., Liermann, N., 2005. Integrated research contributions for screening the tight gas potential in the Rotliegendes formation of North-Germany. OIL GAS European Magazine 121, 187-193.

Reicherter, K., Froitzheim, N., Jarosinski, M., Badura, J., Franzke, H., Hansen, M., Hübscher, C., Müller, R., Poprawa, P., Reinecker, J., 2008. Alpine tectonics north of the Alps, in: McCann, T. (Ed.), The geology of Central Europe. Geological Society of London, pp. 1233-1285.

Reijers, T.J.A., 2012. Sedimentology and diagenesis as 'hydrocarbon exploration tools' in the Late Permian Zechstein-2 Carbonate Member (NE Netherlands). Geologos 18, 163- 195.

Reiter, K., Heidbach, O., Müller, B., Reinecker, J., Röckel, T., 2016. Stress map Germany 2016.

Rider, M.H., Kennedy, M., 2011. The geological interpretation of well logs, 3 ed. Rider and French Consulting Ltd., pp. 440.

Riquelme, A.J., Abellán, A., Tomás, R., Jaboyedoff, M., 2014. A new approach for semi-automatic rock mass joints recognition from 3D point clouds. Computers & Geosciences 68, 38-52.

Ritchie, J., Pratsides, P., 1993. The Caister Fields, Block 44/23a, UK North Sea. Geological Society, London, Petroleum Geology Conference series, pp. 759-769.

Röckel, T., Lempp, C., 2003. Der Spannungszustand im Norddeutschen Becken. Erdöl Erdgas Kohle 119, 73-80.

Rogers, S.F., 2003. Critical stress-related permeability in fractured rocks, in: Ameen, M. (Ed.), Fracture and in-situ stress characterization of hydrocarbon reservoirs. Geological Society, London, Special Publications, pp. 7-16.

Roth, F., Fleckenstein, P., 2001. Stress orientations found in north-east Germany differ from the West European trend. Terra Nova 13, 289-296.

Rygel, M.C., Fielding, C.R., Frank, T.D., Birgenheier, L.P., 2008. The magnitude of Late Paleozoic glacioeustatic fluctuations: A synthesis. Journal of Sedimentary Research 78, 500-511.

Sahoo, H., Gani, M.R., Hampson, G.J., Gani, N.D., Ranson, A., 2016. Facies-to sandbody-scale heterogeneity in a tight-gas fluvial reservoir analog: Blackhawk Formation, Wasatch Plateau, Utah, USA. Marine and Petroleum Geology 78, 48-69.

Sanderson, D.J., Nixon, C.W., 2015. The use of topology in fracture network characterization. Journal of Structural Geology 72, 55-66.

Sannemann, D., Zimdars, J., Plein, E., 1978. Der basale Zechstein (A2-T1) zwischen Weser und Ems. Zeitschrift der deutschen geologischen Gesellschaft, 33-69.

Sathar, S., Reeves, H.J., Cuss, R.J., Harrington, J.F., 2012. The role of stress history on the flow of fluids through fractures. Mineralogical Magazine 76, 3165-3177.

Scheck-Wenderoth, M., Krzywiec, P., Zühlke, R., Maystrenko, Y., Froitzheim, N., 2008. Permian to Cretaceous tectonics of Central Europe, in: McCann, T. (Ed.), The Geology of Central Europe. Geological Society Special Publication, London, pp. 999-1030.

Scheck, M., Bayer, U., 1999. Evolution of the Northeast German Basin - inferences from a 3D structural model and subsidence analysis. Tectonophysics 313, 145-169.

Schmidt, V., McDonald, D.A., 1979. The role of secondary porosity in the course of sandstone diagenesis. SEPM Special Publication 26, 175-207.

Secor, D.T., 1965. Role of fluid pressure in jointing. American Journal of Science 263, 633-646.

Sedat, B., Albertsen, M., Blohm, M., 1992. Petrographie und Diagenese von Sandsteinen im Nordwestdeutschen Oberkarbon. Deutsche Wissenschaftliche Gesellschaft für Erdöl, Erdgas und Kohle (DGMK), Teilbericht 7 zum DGMK-Gemeinschaftsvorhaben 384 Sedimentologie des Oberkarbons, pp. 143.

Seers, T., Hodgetts, D., Wang, Y., Fadlelmula, M., 2017. Direct computation of fracture network equivalent porous medium properties using digital outcrop models. 79th EAGE Conference and Exhibition 2017, Paris, France.

Seers, T.D., Hodgetts, D., 2014. Comparison of digital outcrop and conventional data collection approaches for the characterization of naturally fractured reservoir analogues, in: Spence, G.H., Redfern, J., Aguilera, R., Bevan, T.G., Cosgrove, J.W., Couples, G.D., Daniel, J.-M. (Eds.), Advances in the study of fractured reservoirs. Geological Society, London, Special Publications, pp. 51-77.

Seers, T.D., Hodgetts, D., 2016. Extraction of three-dimensional fracture trace maps from calibrated image sequences. Geosphere 12, 1323-1340.

Segall, P., Fitzgerald, S.D., 1998. A note on induced stress changes in hydrocarbon and geothermal reservoirs. Tectonophysics 289, 117-128.

Senglaub, Y., Brix, M.R., Adriasola, A.C., Littke, R., 2005a. New information on the thermal history of the southwestern Lower Saxony Basin, northern Germany, based on fission track analysis. International Journal of Earth Sciences 94, 876-896.

Senglaub, Y., Littke, R., Brix, M.R., 2005b. Numerical modelling of burial and temperature history as an approach for an alternative interpretation of the Bramsche anomaly, Lower Saxony Basin. International Journal of Earth Sciences 95, 204-224.

Shanley, K.W., Cluff, R.M., Robinson, J.W., 2004. Factors controlling prolific gas production from low-permeability sandstone reservoirs: Implications for resource assessment, prospect development, and risk analysis. AAPG Bulletin 88, 1083-1121.

Shipton, Z., Cowie, P., 2001. Damage zone and slip-surface evolution over μm to km scales in high-porosity Navajo sandstone, Utah. Journal of Structural Geology 23, 1825-1844.

Shipton, Z.K., Cowie, P.A., 2003. A conceptual model for the origin of fault damage zone structures in high-porosity sandstone. Journal of Structural Geology 25, 333-344.

Shipton, Z.K., Soden, A.M., Kirkpatrick, J.D., Bright, A.M., Lunn, R.J., 2006. How thick is a fault? Fault displacement-thickness scaling revisited, in: Abercrombie, R. (Ed.), Earthquakes: Radiated energy and the physics of faulting. AGU, pp. 193-198.

Sibbing, D., Sattler, T., Leibe, B., Kobbelt, L., 2013. Sift-realistic rendering. 3DTV-Conference. IEEE, pp. 56-63.

Sibson, R.H., 2000. Fluid involvement in normal faulting. Journal of Geodynamics 29, 469-499.

Słowakiewicz, M., Tucker, M.E., Pancost, R.D., Perri, E., Mawson, M., 2013. Upper Permian (Zechstein) microbialites: Supratidal through deep subtidal deposition, source rock, and reservoir potential. AAPG Bulletin 97, 1921-1936.

Solano, N., Zambrano, L., Aguilera, R., 2011. Cumulative-gas-production distribution on the Nikanassin tight gas formation, Alberta and British Columbia, Canada SPE Reservoir Evaluation & Engineering 14, 357-376.

Stephen, K.D., Dalrymple, M., 2002. Reservoir simulations developed from an outcrop of incised valley fill strata. AAPG Bulletin 86, 797-822.

Strohmenger, C., Antonini, M., Jager, G., Rockenbauch, K., Strauss, C., 1996. Zechstein 2 carbonate reservoir facies distribution in relation to Zechstein sequence stratigraphy (Upper Permian, Northwest Germany): An integrated approach. Bulletin des Centres de Recherches Exploration-Production Elf Aquitaine 20, 1-35.

Strohmenger, C., Rockenbauch, K., Waldmann, R., 1998. Der Zechstein im Ostdeutschen Raum: Fazies, Diagenese und Reservoirentwicklung des Zechstein 2-Karbonats (Ober-Perm) in Nordostdeutschland, in: Karnin, W.-D., Merkel, D., Piske, J., Schretzenmayr, S. (Eds.), Geowissenschaftliche Ergebnisse der Kohlenwasserstoff-Exploration im Land Brandenburg und im Thüringer Becken in den Jahren 1991-1996 (Zechstein und Rotliegend). Schweizerbart'sche Verlagsbuchhandlung, Geologisches Jahrbuch Reihe A, pp. 81-114.

Strohmenger, C., Strauss, C., 1996. Sedimentology and palynofacies of the Zechstein 2 Carbonate (Upper Permian, Northwest Germany): Implications for sequence stratigraphic subdivision. Sedimentary Geology 102, 55-77.

Stroker, T.M., Harris, N.B., Crawford Elliott, W., Marion Wampler, J., 2013. Diagenesis of a tight gas sand reservoir: Upper Cretaceous Mesaverde Group, Piceance Basin, Colorado. Marine and Petroleum Geology 40, 48-68.

Sturzenegger, M., Stead, D., 2009a. Close-range terrestrial digital photogrammetry and terrestrial laser scanning for discontinuity characterization on rock cuts. Engineering Geology 106, 163-182.

Sturzenegger, M., Stead, D., 2009b. Quantifying discontinuity orientation and persistence on high mountain rock slopes and large landslides using terrestrial remote sensing techniques. Natural Hazards and Earth System Sciences 9, 267-287.

Sturzenegger, M., Yan, M., Stead, D., Elmo, D., 2007. Application and limitations of ground-based laser scanning in rock slope characterization. Proceedings of the first Canadian US rock mechanics symposium, pp. 29-36.

Süss, M., Drozdzewski, G., Schäfer, A., 2000. Sequenzstratigraphie des kohleführenden Oberkarbons im Ruhr-Becken. Geologisches Jahrbuch A 156, 45-106.

Taylor, G.H., Teichmüller, M., Davis, A., Diessel, C.F.K., Littke, R., Robert, P., 1998. Organic Petrology. A New Handbook incorporating some revised parts of Stach's Textbook of Coal Petrology. Berlin, Stuttgart: Gebrüder Borntraeger, pp. 704.

Taylor, J., 1998. Upper Permian - Zechstein, in: Glennie, K. (Ed.), Petroleum Geology of the North Sea: Basic concepts and recent advances. Oxford, United Kingdom, Blackwell Science, pp. 174-210.

Teichmüller, M., 1986. Coalification and natural gas deposits in northwestern Germany, in: Brooks, J., Goff, J.C., van Hoorn, B. (Eds.), Habitat of Paleozoic Gas in N.W. Europe. Geological Society, London, Special Publications, pp. 101-112.

Terzaghi, R.D., 1965. Sources of error in joint surveys. Geotechnique 15, 287-304.

Teufel, L.W., Rhett, D.W., Farrell, H.E., 1991. Effect of reservoir depletion and pore pressure drawdown on in situ stress and deformation in the Ekofisk field, North Sea. The 32nd US Symposium on Rock Mechanics (USRMS). American Rock Mechanics Association.

Thyne, G., 2001. A model for diagenetic mass transfer between adjacent sandstone and shale. Marine and Petroleum Geology 18, 743-755.

Tobin, R.C., McClain, T., Lieber, R.B., Ozkan, A., Banfield, L.A., Marchand, A.M., McRae, L.E., 2010. Reservoir quality modeling of tight-gas sands in Wamsutter field: Integration of diagenesis, petroleum systems, and production data. AAPG Bulletin 94, 1229- 1266.

Torabi, A., Berg, S.S., 2011. Scaling of fault attributes: A review. Marine and Petroleum Geology 28, 1444-1460.

Trendell, A.M., Atchley, S.C., Nordt, L.C., 2012. Depositional and diagenetic controls on reservoir attributes within a fluvial outcrop analog: Upper Triassic Sonsela member of the Chinle Formation, Petrified Forest National Park, Arizona. AAPG Bulletin 96, 679-707.

Tucker, M.E., 1991. Sequence stratigraphy of carbonate-evaporite basins: Models and application to the Upper Permian (Zechstein) of northeast England and adjoining North Sea. Journal of the Geological Society 148, 1019-1036.

Tucker, M.E., 2001. Sedimentary petrology: An introduction to the origin of sedimentary rocks. Wiley-Blackwell, pp. 272.

Van Keer, I., Muchez, P., Viaene, W., 1998. Clay mineralogical variations and evolutions in sandstone sequences near a coal seam and shales in the Westphalian of the Campine Basin (NE Belgium). Clay Minerals 33, 159-169.

Van Wees, J.-D., Stephenson, R., Ziegler, P., Bayer, U., McCann, T., Dadlez, R., Gaupp, R., Narkiewicz, M., Bitzer, F., Scheck, M., 2000. On the origin of the Southern Permian Basin, Central Europe. Marine and Petroleum Geology 17, 43-59.

Walderhaug, O., 1994. Temperatures of quartz cementation in Jurassic sandstones from the Norwegian continental shelf--evidence from fluid inclusions. Journal of Sedimentary Research 64, 311-323.

Walderhaug, O., 2000. Modeling quartz cementation and porosity in Middle Jurassic Brent Group sandstones of the Kvitebjørn field, northern North Sea. AAPG Bulletin 84, 1325- 1339.

Watkins, H., Bond, C.E., Healy, D., Butler, R.W.H., 2015. Appraisal of fracture sampling methods and a new workflow to characterise heterogeneous fracture networks at outcrop. Journal of Structural Geology 72, 67-82.

Weltje, G.J., von Eynatten, H., 2004. Quantitative provenance analysis of sediments: Review and outlook. Sedimentary Geology 171, 1-11.

Wennberg, O., Svånå, T., Azizzadeh, M., Aqrawi, A., Brockbank, P., Lyslo, K., Ogilvie, S., 2006. Fracture intensity vs. mechanical stratigraphy in platform top carbonates: The Aquitanian of the Asmari Formation, Khaviz Anticline, Zagros, SW Iran. Petroleum Geoscience 12, 235-246.

Wennberg, O.P., Casini, G., Jonoud, S., Peacock, D.C.P., 2016. The characteristics of open fractures in carbonate reservoirs and their impact on fluid flow: A discussion. Petroleum Geoscience, 2015-2003.

Wilson, C.E., Aydin, A., Karimi-Fard, M., Durlofsky, L.J., Sagy, A., Brodsky, E.E., Kreylos, O., Kellogg, L.H., 2011. From outcrop to flow simulation: Constructing discrete fracture models from a LIDAR survey. AAPG Bulletin 95, 1883-1905.

Wimmers, K., Koehrer, B., 2014. Integration of sedimentology, petrophysics and rock typing as key to understanding a tight gas reservoir. OIL GAS European Magazine 40, 196- 200.

Wiprut, D., Zoback, M.D., 2002. Fault reactivation, leakage potential, and hydrocarbon column heights in the northern North Sea, in: Koestler, A.G., Hunsdale, R. (Eds.),

Hydrocarbon seal quantification. Elsevier, Norwegian Petroleum Society Special Publications, pp. 203-219.

Worden, R., Burley, S., 2003. Sandstone diagenesis: The evolution of sand to stone, in: Burley, S., Worden, R.H. (Eds.), Sandstone diagenesis: Recent and ancient. Blackwell Publishing, pp. 3-44.

Worden, R., Morad, S., 2003. Clay minerals in sandstones: Controls on formation, distribution and evolution, in: Worden, R.H., Morad, S. (Eds.), Clay mineral cements in sandstones. pp. 3-41.

Wüstefeld, P., 2018. Structure and diagenesis in Upper Carboniferous tight gas reservoirs in NW Germany, PhD thesis. Karlsruhe Institute of Technology, KIT Scientific Publishing, Karlsruhe, Germany, pp. 144.

Wüstefeld, P., De Medeiros, M., Koehrer, B., Sibbing, D., Kobbelt, F., Hilgers, C., 2016. Automated workflow to derive LIDAR fracture statistics for the DFN modelling of a tight gas sandstone reservoir analog, EAGE 78th Conference and Exhibition 2016, Vienna, Austria.

Wüstefeld, P., Hilse, U., Koehrer, B., Adelmann, D., Hilgers, C., 2017a. Critical evaluation of an Upper Carboniferous tight gas sandstone reservoir analog: Diagenesis and petrophysical aspects. Marine and Petroleum Geology 86, 689-710.

Wüstefeld, P., Hilse, U., Lüders, V., Wemmer, K., Koehrer, B., Hilgers, C., 2017b. Kilometer-scale fault-related thermal anomalies in tight gas sandstones. Marine and Petroleum Geology 86, 288-303.

Yale, D.P., 2003. Fault and stress magnitude controls on variations in the orientation of in situ stress, in: Ameen, M. (Ed.), Fracture and in-situ stress characterization of hydrocarbon reservoirs. Geological Society, London, Special Publications, pp. 55-64.

Yurewicz, D., Bohacs, K., Kendall, J., Klimentidis, R., Kronmueller, K., Meurer, M., Ryan, T., Yeakel, J., 2008. Controls on gas and water distribution, Mesaverde basin-centered gas play, Piceance Basin, Colorado, in: Cumella, S., Shanley, K., Camp, W. (Eds.), Understanding, exploring, and developing tight-gas sands. 2005 Vail Hedberg Conference, AAPG Hedberg Series, pp. 105-136.

Zeeb, C., Gomez-Rivas, E., Bons, P.D., Blum, P., 2013. Evaluation of sampling methods for fracture network characterization using outcrops. AAPG Bulletin 97, 1545-1566.

Zeng, L., Tang, X., Jiang, J., Peng, Y., Yang, Y., Lyu, W., 2015. Unreliable determination of in situ stress orientation by borehole breakouts in fractured tight reservoirs: A case study of the upper Eocene Hetaoyuan Formation in the Anpeng field, Nanxiang Basin, China. AAPG Bulletin 99, 1991-2003.

Zhang, Y., Pe-Piper, G., Piper, D.J.W., 2015. How sandstone porosity and permeability vary with diagenetic minerals in the Scotian Basin, offshore eastern Canada: Implications for reservoir quality. Marine and Petroleum Geology 63, 28-45.

Ziegler, K., 2006. Clay minerals of the Permian Rotliegend Group in the North Sea and adjacent areas. Clay Minerals 41, 355-393.

Ziegler, P., 1989. Geodynamic model for Alpine intra-plate compressional deformation in Western and Central Europe, in: Cooper, M.A., Williams, G.D. (Eds.), Inversion tectonics. Geological Society, London, Special Publications, pp. 63-85.

Ziegler, P., 1990. Geological Atlas of Western and Central Europe (2nd edition). Shell Internationale Petroleum Maatschappij BV, Elsevier, Amsterdam.

Zou, C., Zhao, Q., Zhang, G., Xiong, B., 2016. Energy revolution: From a fossil energy era to a new energy era. Natural Gas Industry B 3, 1-11.

Appendix

Table A.1. Sample information of the studied outcrops (Woitzel, Hüggel, Piesberg) and wells (Well A and Well B) containing average point-counted mineral compositions (in %), average point-counted porosities (in %) vs petrophysical measurement data (porosity in %, permeability in mD), grain size data (in mm), and information about the sorting. RF = rock fragments; Mn. Grains = detrital mineral grains.

Woitzel

	Detrital components						**Authigenic components**											**Porosity, permeability, and IGV**						**Grainsize**		
Sample	Quartz	Feldspar	RF	Mica	Matrix	Mn. Grains and Accessories	Quartz cement	Fspar cement	Clay cement	Clay rpl	Ill rpl kaol	Carbonate	Carbonate rpl	Sulphate cements	Pyrite	Fracture fill cement	Fe oxides/-hydroxides	Secondary porosity (intragranular)	Secondary porosity (intergranular)	Fracture porosity	He porosity unstressed	Permeability at 2 MPa	Intergranular Volume (IGV)	Average	Maximum	Sorting
Wo_001	60.3	1.3	14.0	0.0	0.0	0.3	2.0	0.0	11.7	7.3	0.0	0.0	0.0	0.0	0.0	0.0	0.3	0.0	2.7	0.0	14.3	0.0269	16.7	0.3	0.8	moderately sorted
Wo_002	58.3	1.3	16.3	0.0	0.0	0.0	0.0	0.0	4.7	3.3	0.0	0.0	0.0	0.0	0.0	0.0	14.0	0.0	2.0	0.0	13.6	0.0075	20.7	0.3	0.7	moderately sorted
Wo_003	51.0	2.0	22.3	0.0	0.0	0.3	5.0	0.0	8.7	3.0	0.0	0.0	0.0	0.0	0.0	0.0	4.3	0.0	3.3	0.0	15.4	0.0427	21.3	0.4	0.9	moderately sorted
Wo_004	51.3	2.3	21.7	0.7	0.0	0.3	6.7	0.0	6.7	2.3	0.0	0.0	0.0	0.0	0.0	0.0	2.7	0.0	5.3	0.0	13.6	0.0435	21.3	0.4	1.0	moderately sorted
Wo_005	53.7	2.7	19.3	0.0	0.0	1.3	3.0	0.0	10.3	1.0	0.0	0.0	0.0	0.0	0.0	0.0	6.3	0.0	2.3	0.0	15.0	0.0312	22.0	0.4	0.7	moderately sorted
Wo_006	60.3	2.0	18.7	0.0	0.0	1.3	2.3	0.0	7.3	6.0	0.0	0.0	0.0	0.0	0.0	0.0	0.0	0.0	2.0	0.0	15.0	0.0273	11.7	0.4	0.8	moderately sorted
Wo_007	52.3	1.7	22.0	0.3	0.0	0.3	2.3	0.0	6.3	0.7	0.0	0.0	0.0	0.0	0.0	0.0	12.3	0.0	1.7	0.0	14.9	0.0052	22.7	0.4	0.9	moderately sorted
Wo_008	54.0	2.7	15.7	0.3	0.0	0.3	4.0	0.0	8.0	1.0	0.0	0.0	0.0	0.0	0.0	0.0	6.3	0.0	7.7	0.0	13.9	0.0048	26.0	0.5	0.9	moderately sorted
Wo_009	60.7	0.3	15.7	0.0	0.0	0.7	4.7	0.0	4.7	2.7	0.0	0.0	0.0	0.0	0.0	0.0	9.0	0.0	1.7	0.0	13.6	0.0051	20.0	0.5	1.0	moderately sorted
Wo_010	61.7	1.3	15.3	0.0	0.0	1.7	2.7	0.0	7.0	6.3	0.0	0.0	0.0	0.0	0.0	0.0	1.0	0.0	3.0	0.0	15.3	0.0482	13.7	0.4	0.7	moderately sorted
Wo_011	55.7	1.3	19.7	0.0	0.0	1.3	3.0	0.0	7.3	2.3	0.0	0.0	0.0	0.0	0.0	0.0	5.3	0.0	4.0	0.0	15.8	0.0355	19.7	0.4	1.1	moderately sorted
Wo_012	57.0	2.0	19.7	0.0	0.0	1.3	4.3	0.0	7.7	3.7	0.0	0.0	0.0	0.0	0.0	0.0	3.3	0.0	1.0	0.0	15.2	0.0678	16.3	0.4	0.8	moderately sorted
Wo_013	54.7	1.7	22.3	0.0	0.0	1.0	4.7	0.0	2.3	1.0	0.0	0.0	0.0	0.0	0.0	0.0	7.7	0.0	4.7	0.0	13.9	0.0535	19.3	0.5	0.9	moderately sorted
Wo_015	53.7	0.3	22.0	0.3	0.0	0.7	4.7	0.0	5.7	2.3	0.0	0.0	0.0	0.0	0.0	0.0	7.7	0.0	2.7	0.0	14.3	0.0032	20.7	0.5	1.7	moderately well sorted
Wo_018	59.0	1.7	16.0	0.0	0.0	1.0	4.3	0.0	10.7	2.7	0.0	0.0	0.0	0.0	0.0	0.0	0.0	0.0	4.7	0.0	16.3	0.2000	19.7	0.6	1.5	moderately well sorted
Wo_020	53.3	1.7	22.3	0.0	0.0	0.0	2.7	0.0	4.7	2.7	0.0	0.0	0.0	0.0	0.0	0.0	8.7	0.0	4.0	0.0	15.6	0.0509	20.0	0.4	0.8	moderately sorted
Wo_021	54.3	1.0	20.0	0.0	0.0	2.0	2.7	0.0	11.3	2.7	0.0	0.0	0.0	0.0	0.0	0.0	0.3	0.0	5.7	0.0	19.0	0.1260	20.0	0.5	2.4	moderately sorted
Wo_022	45.7	1.7	28.7	0.7	0.0	2.0	1.7	0.0	12.3	3.7	0.0	0.0	0.0	0.0	0.0	0.0	0.3	0.0	3.3	0.0	18.5	0.0177	17.7	0.3	0.5	moderately sorted
Wo_023	38.0	2.3	36.0	2.3	0.0	1.3	1.7	0.0	3.0	2.3	0.0	0.0	0.0	0.0	0.0	0.0	12.0	0.0	1.0	0.0	17.2	0.0379	17.7	0.3	0.6	moderately sorted
Wo_025	46.3	0.0	34.3	0.3	0.0	1.3	4.3	0.0	1.3	2.7	0.0	0.0	0.0	0.0	0.0	0.0	9.0	0.0	0.3	0.0	15.0	0.0118	15.0	0.6	1.1	moderately well sorted
Wo_026	59.7	1.7	15.7	0.0	0.0	0.3	5.0	0.0	7.7	4.3	0.0	0.0	0.0	0.0	0.0	0.0	2.3	0.0	3.3	0.0	16.8	0.0657	18.3	0.4	0.8	moderately sorted
Wo_027	59.0	2.3	19.0	0.0	0.0	0.7	5.0	0.0	6.7	2.7	0.0	0.0	0.0	0.0	0.0	0.0	0.0	0.0	4.7	0.0	22.7	0.0429	16.3	0.4	0.8	moderately sorted
Wo_028	47.7	2.3	27.3	0.0	0.0	1.7	2.7	0.0	9.7	4.0	0.0	0.0	0.0	0.0	0.0	0.0	0.0	0.0	4.7	0.0	16.5	0.0365	17.0	0.5	1.0	moderately sorted
Wo_029	52.3	2.3	24.0	0.0	0.0	1.3	4.7	0.0	9.3	3.7	0.0	0.0	0.0	0.0	0.0	0.0	0.3	0.0	2.0	0.0	16.7	0.0239	16.3	0.4	1.0	moderately sorted
Wo_030	54.0	2.7	19.3	0.0	0.0	0.7	7.7	0.0	4.0	1.0	0.0	0.0	0.0	0.0	0.0	0.0	4.7	0.0	6.0	0.0	16.4	0.0169	22.3	0.4	0.8	moderately sorted
Wo_031	52.0	1.7	19.3	1.3	0.0	0.3	1.3	0.0	4.0	1.7	0.0	0.0	0.0	0.0	0.0	0.0	14.3	0.0	4.0	0.0	16.2	0.0195	23.7	0.4	1.0	moderately sorted
Wo_033	48.7	1.0	25.3	0.3	0.0	0.0	5.0	0.0	2.7	3.0	0.0	0.0	0.0	0.0	0.0	0.0	13.3	0.0	0.7	0.0	16.8	0.0017	21.7	0.4	0.9	moderately sorted
Wo_034	51.7	1.3	19.3	0.3	0.0	0.3	4.7	0.0	7.0	2.3	0.0	0.0	0.0	0.0	0.0	0.0	9.3	0.0	3.7	0.0	15.2	0.0280	24.7	0.4	1.2	moderately sorted
Wo_035	47.3	1.3	22.7	0.0	0.0	3.0	5.3	0.0	5.0	4.3	0.0	0.0	0.0	0.0	0.0	0.0	4.3	0.0	6.7	0.0	14.7	0.0599	21.3	0.5	1.1	moderately well sorted
Average	**53.6**	**1.7**	**21.2**	**0.2**	**0.0**	**0.9**	**3.7**	**0.0**	**6.8**	**3.0**	**0.0**	**0.0**	**0.0**	**0.0**	**0.0**	**0.0**	**5.5**	**0.0**	**3.4**	**0.0**	**15.8**	**0.04**	**19.4**	**0.4**	**1.0**	
Max	**61.7**	**2.7**	**36.0**	**2.3**	**0.0**	**3.0**	**7.7**	**0.0**	**12.3**	**7.3**	**0.0**	**0.0**	**0.0**	**0.0**	**0.0**	**0.0**	**14.3**	**0.0**	**7.7**	**0.0**	**22.7**	**0.2**	**26.0**	**0.6**	**2.4**	
Min	**38.0**	**0.0**	**14.0**	**0.0**	**0.0**	**0.0**	**0.0**	**0.0**	**1.3**	**0.7**	**0.0**	**0.0**	**0.0**	**0.0**	**0.0**	**0.0**	**0.0**	**0.0**	**0.3**	**0.0**	**13.6**	**0.0**	**11.7**	**0.3**	**0.5**	

Table A.1 (continued)

		Detrital components						Authigenic components											Porosity, permeability, and IGV						Grainsize		
	Sample	Quartz	Feldspar	RF	Mica	Matrix	Mn. Grains and Accessories	Quartz cement	Fspar cement	Clay cement	Clay rpl	Ill rpl kaol	Carbonate	Carbonate rpl	Sulphate cements	Pyrite	Fracture fill cement	Fe oxides/-hydroxides	Secondary porosity (intragranular)	Secondary porosity (intergranular)	Fracture porosity	He porosity unstressed	Permeability at 2 MPa	Intergranular Volume (IGV)	Average	Maximum	Sorting
Hüggel	Hu_02	50.0	0.0	21.0	0.0	0.0	1.0	5.0	0.0	5.3	4.0	0.0	0.0	0.0	0.0	0.0	0.0	1.3	0.0	12.3	0.0	-	-	24.0	0.4	0.7	moderately sorted
	Hu_03	46.0	0.0	25.3	0.3	0.0	0.0	4.3	0.0	2.3	1.0	0.0	8.3	0.0	0.0	0.0	0.0	5.7	0.0	6.7	0.0	-	-	27.3	0.4	0.8	moderately sorted
	Hu_04	44.3	0.0	22.0	0.0	0.0	0.3	5.3	0.0	4.7	1.7	0.0	0.0	0.0	0.0	0.0	0.0	11.3	0.0	10.3	0.0	-	-	31.7	0.4	0.8	moderately sorted
	Hu_05	52.0	0.0	18.3	0.0	0.0	0.3	2.0	0.0	6.0	0.3	0.0	15.7	0.0	0.0	0.0	0.0	1.7	0.0	3.7	0.0	30.9	0.3529	29.0	0.4	0.9	moderately sorted
	Hu_06	39.7	0.0	47.0	0.0	0.0	0.7	0.7	0.0	5.3	0.7	0.0	0.0	0.0	0.0	0.0	0.0	6.0	0.0	0.0	0.0	15.5	0.0182	12.0	0.2	0.3	moderately sorted
	Hu_07	46.7	0.0	27.0	0.0	0.0	4.3	0.3	0.0	8.7	0.7	0.0	0.0	0.0	0.0	0.0	0.0	11.3	0.0	1.0	0.0	17.3	0.0394	21.3	0.3	0.5	moderately sorted
	Hu_08	61.0	1.0	17.7	0.0	0.0	0.3	3.0	0.0	7.7	2.7	0.7	0.3	0.0	0.0	0.0	0.0	0.7	0.0	5.0	0.0	20.4	2.4500	16.7	0.6	1.4	moderately sorted
	Hu_11	55.7	2.3	22.3	0.3	0.0	0.0	1.3	0.0	8.0	2.0	0.7	0.0	0.0	0.0	0.0	0.0	2.7	0.0	4.7	0.0	18.6	3.1334	16.7	0.5	2.2	moderately well sorted
	Hu_12	46.7	0.7	29.3	1.3	0.0	0.3	1.0	0.0	11.7	2.0	0.7	0.0	0.0	0.0	0.0	0.0	5.3	0.0	1.0	0.0	15.6	0.0495	19.0	0.2	0.6	moderately sorted
	Hu_15	57.7	1.0	20.3	0.7	0.0	0.0	2.7	0.0	8.3	0.7	1.3	0.3	0.0	0.0	0.0	0.0	4.0	0.0	3.0	0.0	15.1	0.0279	18.3	0.4	1.0	moderately sorted
	Hu_17	53.0	0.3	17.7	0.0	0.0	3.3	2.7	0.0	4.7	0.7	0.0	0.0	0.0	0.0	0.0	0.0	15.0	0.0	2.7	0.0	16.8	0.2824	25.0	0.3	0.8	moderately sorted
	Average	**50.2**	**0.5**	**24.4**	**0.2**	**0.0**	**1.0**	**2.6**	**0.0**	**6.6**	**1.5**	**0.3**	**2.2**	**0.0**	**0.0**	**0.0**	**0.0**	**5.9**	**0.0**	**4.6**	**0.0**	**18.8**	**0.8**	**21.9**	**0.4**	**0.9**	
	Max	**61.0**	**2.3**	**47.0**	**1.3**	**0.0**	**4.3**	**5.3**	**0.0**	**11.7**	**4.0**	**1.3**	**15.7**	**0.0**	**0.0**	**0.0**	**0.0**	**15.0**	**0.0**	**12.3**	**0.0**	**30.9**	**3.1**	**31.7**	**0.6**	**2.2**	
	Min	**39.7**	**0.0**	**17.7**	**0.0**	**0.0**	**0.0**	**0.3**	**0.0**	**2.3**	**0.3**	**0.0**	**0.0**	**0.0**	**0.0**	**0.0**	**0.0**	**0.7**	**0.0**	**0.0**	**0.0**	**15.1**	**0.0**	**12.0**	**0.2**	**0.3**	
Piesberg	SS02	49.7	1.7	8.0	0.3	0.0	0.0	7.7	0.0	3.0	0.0	4.0	0.0	0.0	0.0	0.0	0.0	7.0	0.0	18.7	0.0	15.7	0.0084	36.3	0.4	0.7	moderately sorted
	SS03	59.3	0.3	12.7	0.0	0.0	0.0	17.7	0.0	4.3	0.0	3.3	0.0	0.0	0.0	0.0	0.0	1.7	0.0	0.7	0.0	7.2	0.0005	24.3	0.5	0.9	moderately sorted
	SS05	74.3	2.7	14.7	0.0	0.0	0.0	1.3	0.0	0.7	0.0	0.3	0.0	0.0	0.0	0.0	0.0	2.3	0.0	3.7	0.0	8.3	0.0004	8.0	0.4	1.1	moderately sorted
	SS09	64.0	0.0	12.7	0.0	0.0	0.0	10.3	0.0	3.0	0.0	0.0	0.0	0.0	0.0	0.0	0.0	3.3	0.0	6.7	0.0	10.3	0.0017	23.3	0.4	1.1	moderately sorted
	SS13	61.7	0.7	18.0	0.7	0.0	0.0	8.7	0.0	3.0	0.3	0.7	0.0	0.0	0.0	0.0	0.0	2.3	0.0	4.0	0.0	5.9	0.0020	18.0	0.5	1.0	moderately sorted
	SS16	50.0	2.3	42.0	1.3	0.0	0.0	1.3	0.0	2.0	1.0	0.0	0.0	0.0	0.0	0.0	0.0	0.0	0.0	0.0	0.0	0.2	0.0000	3.3	0.3	0.6	moderately sorted
	SS18	68.3	0.7	16.0	0.0	0.0	0.0	7.0	0.0	3.7	0.0	1.3	0.0	0.0	0.0	0.0	0.0	2.0	0.0	1.0	0.0	8.6	0.0008	13.7	0.5	0.8	moderately sorted
	SS22a	53.7	0.3	18.0	0.3	0.0	0.0	15.0	0.0	6.0	0.0	1.3	0.0	0.0	0.0	0.0	0.0	2.3	0.0	3.0	0.0	6.6	0.0002	26.3	0.5	0.9	moderately sorted
	SS23	62.3	0.3	20.3	0.0	0.0	0.0	10.3	0.0	1.7	0.0	0.7	0.0	0.0	0.0	0.0	0.0	2.7	0.0	1.7	0.0	6.6	0.0003	16.3	0.6	1.3	moderately well sorted
	SS25	78.0	1.7	9.7	0.7	0.0	0.0	2.0	0.0	1.3	0.0	0.7	0.0	0.0	0.0	0.0	0.0	5.0	0.0	1.0	0.0	7.2	-	9.3	0.5	0.9	moderately sorted
	SS26	59.7	0.3	8.0	0.0	0.0	0.0	8.7	0.0	11.0	0.0	0.3	0.0	0.0	0.0	0.0	0.0	7.3	0.0	4.7	0.0	11.6	0.0001	31.7	0.3	0.6	moderately sorted
	SS27	59.0	1.3	16.0	0.0	0.0	0.0	10.0	0.0	4.0	0.0	1.0	0.0	0.0	0.0	0.0	0.0	0.0	0.0	8.7	0.0	11.5	0.0008	22.7	0.4	0.9	moderately sorted
	SS28	53.0	1.0	17.3	0.3	0.0	0.0	6.3	0.0	3.3	1.0	1.7	0.0	0.0	0.0	0.0	0.0	8.3	0.0	7.7	0.0	14.7	0.0607	25.7	0.5	2.3	moderately well sorted
	SS33	49.3	0.7	28.7	0.3	0.0	0.0	3.3	0.0	6.7	0.0	2.3	0.0	0.0	0.0	0.0	0.0	2.7	0.0	6.0	0.0	9.6	-	18.7	0.3	1.2	moderately sorted
	MK03	56.0	0.3	17.3	0.7	0.0	0.0	12.0	0.0	4.3	0.0	2.3	0.0	0.0	0.0	0.0	0.0	3.3	0.0	3.7	0.0	4.6	0.0001	23.3	0.3	0.5	moderately sorted
	MK04a	67.7	2.7	15.0	1.0	0.0	0.0	2.0	0.0	0.7	0.0	0.0	0.0	0.0	0.0	0.0	0.0	8.3	0.0	2.7	0.0	8.7	0.0009	13.7	0.3	0.6	moderately sorted
	MK05	48.0	0.3	9.7	0.0	0.0	0.0	6.7	0.0	4.0	0.0	1.7	0.0	0.0	0.0	0.0	0.0	11.3	0.0	18.3	0.0	19.9	-	40.3	0.3	0.7	moderately sorted
	MK06	55.3	1.0	12.3	0.0	0.0	0.0	15.3	0.0	5.3	1.0	2.0	0.0	0.0	0.0	0.0	0.0	3.7	0.0	4.0	0.0	7.0	0.0011	28.3	0.3	0.5	moderately sorted
	MK08	78.0	1.0	14.3	0.7	0.0	0.0	2.3	0.0	0.7	0.0	0.0	0.0	0.0	0.0	0.0	0.0	2.3	0.0	0.7	0.0	6.2	0.0008	6.0	0.4	0.8	moderately sorted

Table A.1 (continued)

Piesberg

	Detrital components						Authigenic components											Porosity, permeability, and IGV						Grainsize		
Sample	Quartz	Feldspar	RF	Mica	Matrix	Mn. Grains and Accessories	Quartz cement	Fspar cement	Clay cement	Clay rpl	Ill rpl kaol	Carbonate	Carbonate rpl	Sulphate cements	Pyrite	Fracture fill cement	Fe oxides/-hydroxides	Secondary porosity (intragranular)	Secondary porosity (intergranular)	Fracture porosity	He porosity unstressed	Permeability at 2 MPa	Intergranular Volume (IGV)	Average	Maximum	Sorting
MK09	60.0	0.7	13.7	0.7	0.0	0.0	14.3	0.0	5.0	0.0	1.3	0.0	0.0	0.0	0.0	0.0	3.7	0.0	0.7	0.0	6.5	0.0006	23.7	0.3	0.6	moderately sorted
MK11	56.7	0.7	18.0	0.0	0.0	0.0	11.0	0.0	5.7	0.0	2.3	0.0	0.0	0.0	0.0	0.0	2.0	0.0	3.7	0.0	6.0	0.0010	22.3	0.4	0.7	moderately sorted
MK12	57.0	0.3	14.7	0.0	0.0	0.0	15.0	0.0	5.3	0.0	3.7	0.0	0.0	0.0	0.0	0.0	1.7	0.0	2.3	0.0	5.0	0.0008	24.3	0.4	0.9	moderately sorted
MK13	64.3	0.3	10.3	0.0	0.0	0.0	10.3	0.0	8.0	0.0	1.3	0.0	0.0	0.0	0.0	0.0	2.7	0.0	2.7	0.0	6.0	0.0003	23.7	0.3	0.7	moderately sorted
MK14	61.0	0.0	16.7	0.0	0.0	0.0	15.3	0.0	3.3	0.0	2.3	0.0	0.0	0.0	0.0	0.0	1.3	0.0	0.0	0.0	8.1	0.0006	20.0	0.4	0.8	moderately sorted
MK16	54.0	1.3	19.0	0.3	0.0	0.0	19.0	0.0	2.0	0.0	2.7	0.7	0.0	0.0	0.0	0.0	0.3	0.0	0.7	0.0	5.2	-	22.7	0.4	0.7	moderately sorted
MK17	63.3	0.7	11.7	0.7	0.0	0.0	11.7	0.0	6.0	0.0	2.0	0.0	0.0	0.0	0.0	0.0	1.7	0.0	2.3	0.0	5.7	0.0004	21.7	0.4	0.8	moderately sorted
MK24	55.0	0.0	11.3	0.3	0.0	0.0	23.7	0.0	1.7	0.0	3.7	0.0	0.0	0.0	0.0	0.0	3.7	0.0	0.7	0.0	5.0	0.0007	29.7	0.5	0.9	moderately sorted
PB008	57.7	1.3	17.3	0.0	0.0	0.0	15.7	0.0	6.7	0.0	1.3	0.0	0.0	0.0	0.0	0.0	0.0	0.0	0.0	0.0	1.8	0.0002	22.3	0.4	0.7	moderately sorted
PB034	52.7	0.7	16.7	0.0	0.0	0.0	17.0	0.0	6.7	0.0	2.0	0.0	0.0	0.0	0.0	0.0	1.7	0.0	2.7	0.0	8.7	0.0004	28.0	0.4	0.9	moderately sorted
PB009	58.3	0.3	12.3	0.0	0.0	0.0	19.3	0.0	5.3	0.0	2.0	0.0	0.0	0.0	0.0	0.0	0.0	0.0	2.3	0.0	6.0	-	27.0	0.6	1.0	moderately sorted
PB019	59.0	0.7	9.7	0.0	0.0	0.0	22.3	0.0	5.3	0.0	2.0	0.0	0.0	0.0	0.0	0.0	1.0	0.0	0.0	0.0	4.3	0.0014	28.7	0.5	1.0	moderately sorted
PB018	65.0	0.3	6.3	0.3	0.0	0.0	18.3	0.0	5.3	0.0	2.0	0.0	0.0	0.0	0.0	0.0	0.7	0.0	1.7	0.0	-	-	26.0	0.5	1.1	moderately sorted
PB006	55.3	0.3	12.3	0.0	0.0	0.0	25.0	0.0	5.3	0.0	1.0	0.0	0.0	0.0	0.0	0.0	0.7	0.0	0.0	0.0	2.8	-	31.0	0.6	1.5	moderately well sorted
PB004	63.3	0.7	8.0	0.0	0.0	0.0	21.3	0.0	3.3	0.0	1.0	0.0	0.0	0.0	0.0	0.0	0.0	0.0	2.3	0.0	3.3	-	27.0	0.6	1.8	moderately well sorted
PB014	48.3	0.7	28.0	1.0	0.0	0.0	11.3	0.0	3.0	0.0	3.7	0.0	0.0	0.0	0.0	0.0	3.0	0.0	1.0	0.0	0.5	0.0002	18.3	0.3	0.5	moderately sorted
PB068	54.7	2.0	14.3	0.3	0.0	0.0	7.3	0.0	3.7	0.0	2.0	0.0	0.0	0.0	0.0	0.0	5.3	0.0	10.3	0.0	16.5	-	26.7	0.4	1.1	moderately sorted
PB069	55.3	0.7	13.0	0.0	0.0	0.0	17.7	0.0	9.0	0.0	0.3	0.0	0.0	0.0	0.0	0.0	0.0	0.0	4.0	0.0	7.2	0.0006	30.7	0.4	2.0	moderately sorted
PB073	64.0	0.7	15.0	0.7	0.0	0.0	14.0	0.0	4.3	0.0	0.7	0.0	0.0	0.0	0.0	0.0	0.0	0.0	0.7	0.0	4.2	0.0001	19.0	0.3	0.6	moderately sorted
PB075	55.7	1.0	21.3	0.7	0.0	0.0	14.3	0.0	3.7	0.0	0.3	0.0	0.0	0.0	0.0	0.0	0.7	0.0	2.3	0.0	5.8	0.0001	21.0	0.3	0.8	moderately sorted
PB080	53.7	0.7	13.3	0.3	0.0	0.0	17.7	0.0	6.0	0.0	2.3	0.0	0.0	0.0	0.0	0.0	4.3	0.0	1.7	0.0	13.1	0.0001	29.7	0.4	0.7	moderately sorted
PB011	58.3	0.7	7.3	0.0	0.0	0.0	18.3	0.0	3.3	0.0	3.0	0.0	0.0	0.0	0.0	0.0	5.7	0.0	3.3	0.0	7.3	-	30.7	0.5	1.2	moderately sorted
PB013	56.0	0.7	16.3	0.3	0.0	0.0	19.7	0.0	3.0	0.0	2.0	0.0	0.0	0.0	0.0	0.0	0.3	0.0	1.7	0.0	4.0	-	24.7	0.4	0.8	moderately sorted
PB098	57.0	0.7	11.0	0.0	0.0	0.0	11.7	0.0	0.3	0.0	0.0	0.0	0.0	0.0	0.0	0.0	0.0	0.0	19.3	0.0	-	-	31.3	0.4	0.8	moderately sorted
L2.0	54.3	0.3	14.0	0.0	0.0	0.0	20.7	0.0	4.3	0.0	4.3	0.0	0.0	0.0	0.0	0.0	0.0	0.0	2.0	0.0	8.8	-	27.0	0.4	0.8	moderately sorted
R0.75	60.7	1.7	19.0	4.3	0.0	0.0	5.3	0.0	2.3	0.0	0.0	0.0	0.0	0.0	0.0	0.0	4.7	0.0	2.0	0.0	11.9	-	14.3	0.3	0.7	moderately sorted
R1.0	57.7	0.0	17.7	0.0	0.0	0.0	18.7	0.0	3.3	0.0	1.3	0.0	0.0	0.0	0.0	0.0	1.0	0.0	0.3	0.0	1.0	-	23.3	0.4	1.1	moderately sorted
Average	**58.8**	**0.8**	**15.2**	**0.4**	**0.0**	**0.0**	**12.7**	**0.0**	**4.1**	**0.1**	**1.6**	**0.0**	**0.0**	**0.0**	**0.0**	**0.0**	**2.7**	**0.0**	**3.6**	**0.0**	**7.4**	**0.003**	**23.1**	**0.4**	**0.9**	
Max	**78.0**	**2.7**	**42.0**	**4.3**	**0.0**	**0.0**	**25.0**	**0.0**	**11.0**	**1.0**	**4.3**	**0.7**	**0.0**	**0.0**	**0.0**	**0.0**	**11.3**	**0.0**	**19.3**	**0.0**	**19.9**	**0.1**	**40.3**	**0.6**	**2.3**	
Min	**48.0**	**0.0**	**6.3**	**0.0**	**0.0**	**0.0**	**1.3**	**0.0**	**0.3**	**0.0**	**0.0**	**0.0**	**0.0**	**0.0**	**0.0**	**0.0**	**0.0**	**0.0**	**0.0**	**0.0**	**0.2**	**0.0**	**3.3**	**0.3**	**0.5**	

Table A.1 (continued)

		Detrital components						Authigenic components										Porosity, permeability, and IGV						Grainsize		
	Sample	Quartz	Feldspar	RF	Mica	Matrix	Mn. Grains and Accessories	Quartz cement	Fspar cement	Clay cement	Clay rpl	Ill rpl kaol	Carbonate	Carbonate rpl	Sulphate cements	Pyrite	Fracture fill cement	Fe oxides/-hydroxides	Secondary porosity (intragranular)	Secondary porosity (intergranular)	Fracture porosity	He porosity unstressed	Permeability at 2 MPa	Intergranular Volume (IGV)	Average	Maximum	Sorting
high porous and permeable Piesberg	PB001	42.7	0.7	29.0	1.0	0.0	0.0	11.0	0.0	10.3	0.0	0.0	0.0	0.0	0.0	0.0	0.0	0.0	0.0	5.3	0.0	14.5	0.4848	26.7	0.1	0.2	moderately sorted
	R20.0	43.7	1.0	15.0	0.3	0.0	0.0	18.7	0.0	7.7	0.0	1.7	0.0	0.0	0.0	0.0	0.0	0.0	0.0	12.0	0.0	15.1	1.3806	38.3	0.3	0.6	moderately sorted
	PB099	48.0	0.0	13.7	0.0	0.0	0.0	11.7	0.0	0.0	0.0	0.0	0.0	0.0	0.0	0.0	0.0	0.0	0.0	26.7	0.0	26.3	219.1329	38.3	0.4	0.8	moderately sorted
	PB100	42.0	0.7	20.3	0.3	0.0	0.0	14.3	0.0	10.3	0.7	1.7	0.0	0.0	0.0	0.0	0.0	0.0	0.0	9.7	0.0	14.4	0.3139	34.3	0.3	0.7	moderately sorted
	R2.0	52.3	0.3	14.3	0.0	0.0	0.0	6.0	0.0	7.0	0.0	1.7	0.0	0.0	0.0	0.0	0.0	5.0	0.0	13.3	0.0	23.6	52.1775	31.3	0.3	0.8	moderately sorted
	Average	**45.7**	**0.5**	**18.5**	**0.3**	**0.0**	**0.0**	**12.3**	**0.0**	**7.1**	**0.1**	**1.0**	**0.0**	**0.0**	**0.0**	**0.0**	**0.0**	**1.0**	**0.0**	**13.4**	**0.0**	**18.8**	**54.7**	**33.8**	**0.3**	**0.6**	
	Max	**52.3**	**1.0**	**29.0**	**1.0**	**0.0**	**0.0**	**18.7**	**0.0**	**10.3**	**0.7**	**1.7**	**0.0**	**0.0**	**0.0**	**0.0**	**0.0**	**5.0**	**0.0**	**26.7**	**0.0**	**26.3**	**219.1**	**38.3**	**0.4**	**0.8**	
	Min	**42.0**	**0.0**	**13.7**	**0.0**	**0.0**	**0.0**	**6.0**	**0.0**	**0.0**	**0.0**	**0.0**	**0.0**	**0.0**	**0.0**	**0.0**	**0.0**	**0.0**	**0.0**	**5.3**	**0.0**	**14.4**	**0.3**	**26.7**	**0.1**	**0.2**	
Well A	A_390183	57.0	0.0	18.0	1.3	1.3	0.0	9.0	0.0	3.0	1.7	1.0	4.3	0.3	0.0	0.0	0.0	0.0	2.3	0.7	0.0	5.9	0.8154	18.3	0.4	0.8	moderately sorted
	A_390249	52.3	1.7	12.7	0.7	0.0	0.0	9.0	0.0	6.0	2.3	4.3	8.7	0.3	0.0	0.0	0.0	0.0	1.7	0.3	0.0	6.8	0.1023	24.0	0.3	0.7	moderately sorted
	A_390356	55.7	0.3	12.0	2.0	0.0	0.0	4.3	0.3	4.3	0.3	3.3	16.0	0.0	0.0	0.0	0.0	0.7	0.3	0.3	0.0	2.9	0.0173	26.0	0.3	0.9	moderately sorted
	A_390429	54.0	0.7	19.7	0.3	0.0	0.3	8.3	0.0	5.0	4.3	4.3	2.3	0.0	0.0	0.0	0.0	0.7	0.0	0.0	0.0	4.4	0.0123	16.3	0.2	0.5	moderately sorted
	A_390521	53.0	0.0	18.0	2.3	0.0	0.3	9.3	0.0	3.0	0.7	3.0	8.3	0.0	0.0	0.0	0.0	0.0	2.0	0.0	0.0	6.4	0.0769	20.7	0.2	0.4	moderately sorted
	A_390604	50.3	0.0	14.3	0.3	0.0	0.3	6.3	0.0	2.0	0.3	1.3	23.3	0.3	0.0	0.0	0.0	0.3	0.7	0.0	0.0	3.5	0.0268	32.0	0.2	0.4	moderately sorted
	A_390758	46.3	0.7	19.3	3.0	0.0	0.0	10.7	0.0	4.0	2.7	2.0	7.7	0.3	0.0	0.0	0.0	0.3	2.0	1.0	0.0	6.2	0.0962	23.7	0.2	0.4	moderately sorted
	A_390865	47.0	0.0	20.7	1.3	0.0	0.0	10.7	0.3	5.7	3.0	2.7	7.7	0.0	0.0	0.0	0.0	0.7	0.3	0.0	0.0	4.3	0.0635	25.0	0.2	0.4	moderately sorted
	A_391104	49.0	0.7	16.3	2.0	0.0	0.0	12.0	0.0	3.0	5.3	3.0	6.3	0.7	0.0	0.0	0.0	0.0	1.0	0.7	0.0	6.0	0.0949	22.0	0.2	0.7	moderately sorted
	A_391161	47.3	0.0	11.3	2.0	0.0	0.3	17.7	0.0	2.7	4.0	2.3	8.3	0.3	0.0	0.0	0.0	1.0	2.0	0.7	0.0	5.6	0.1201	30.3	0.3	0.8	moderately sorted
	A_391338	51.0	0.7	10.7	1.3	0.3	0.0	14.3	0.3	2.0	5.7	2.3	7.0	0.3	0.0	0.0	0.0	0.3	3.0	0.7	0.0	6.2	0.1436	25.0	0.3	0.8	moderately sorted
	A_391573	50.7	0.7	13.3	0.3	0.7	0.0	14.3	0.3	2.7	7.3	2.3	3.3	0.3	0.7	0.0	0.0	0.3	2.0	0.7	0.0	7.3	0.8205	23.0	0.4	0.9	moderately sorted
	A_391875	54.7	0.3	9.7	0.3	0.3	0.0	18.7	0.0	1.0	6.0	2.7	3.7	0.3	0.3	0.0	0.0	0.0	1.3	0.7	0.0	8.9	0.8047	24.7	0.4	1.5	moderately sorted
	A_392100	57.3	0.0	8.7	1.0	0.0	0.0	14.0	0.0	1.3	5.7	1.3	6.3	0.0	0.3	0.0	0.0	0.0	3.3	0.7	0.0	7.9	0.3801	22.7	0.3	0.6	moderately sorted
	A_392397	52.3	1.0	10.3	0.0	0.0	0.0	18.7	0.0	6.0	2.7	2.3	3.7	0.0	0.0	0.0	0.0	0.3	1.7	1.0	0.0	7.2	0.1934	29.7	0.3	0.6	moderately sorted
	A_392586	50.3	1.3	11.3	0.3	0.0	0.0	20.0	0.0	4.3	2.3	1.3	4.3	0.3	0.3	0.0	0.0	0.3	3.0	0.3	0.0	8.4	0.1931	29.7	0.3	0.5	moderately sorted
	A_392673	52.3	0.7	13.7	0.3	0.0	0.0	16.7	0.0	4.0	3.3	2.0	3.7	0.0	0.0	0.0	0.0	0.3	2.3	0.7	0.0	6.1	0.1139	25.3	0.2	0.5	moderately sorted
	A_392789	44.0	0.0	17.3	0.3	0.0	0.0	19.0	0.0	2.7	6.0	1.3	5.0	0.3	0.0	0.0	0.0	0.0	2.7	1.3	0.0	7.7	0.2287	28.0	0.3	0.5	moderately sorted
	A_399858	48.7	0.7	18.7	1.3	0.0	0.0	14.3	0.0	3.7	3.0	2.3	5.0	0.3	0.0	0.0	0.0	0.7	0.7	0.7	0.0	6.5	0.1879	24.3	0.2	0.4	moderately sorted
	A_399986	48.7	0.7	20.3	0.7	0.0	0.3	15.0	0.0	2.7	3.3	0.3	4.7	0.3	0.0	0.0	0.0	1.0	1.7	0.3	0.0	5.4	0.1312	23.7	0.3	0.5	moderately sorted
	A_400217	47.0	0.7	17.7	1.3	0.0	0.0	15.0	0.0	2.7	5.0	3.7	3.7	0.3	0.0	0.0	0.0	0.3	1.3	1.3	0.0	8.6	0.2493	23.0	0.3	0.7	moderately sorted
	A_400373	57.0	0.7	17.3	1.0	0.0	0.0	11.0	0.0	3.7	3.3	1.0	2.0	0.0	0.0	0.0	0.0	0.3	1.7	1.0	0.0	6.3	0.1177	18.0	0.3	0.8	moderately sorted
	A_400435	51.7	0.3	17.7	1.7	0.3	0.0	12.0	0.0	2.3	3.0	2.3	3.7	0.3	0.0	0.0	0.0	0.7	3.3	0.7	0.0	7.3	0.2065	19.7	0.3	0.8	moderately sorted
	A_400526	55.0	0.7	16.0	1.0	0.0	1.0	17.0	0.0	3.0	2.7	1.3	1.0	0.3	0.0	0.0	0.0	0.0	1.0	0.0	0.0	6.6	0.1312	21.0	0.3	0.5	moderately sorted
	A_400685	51.7	0.7	17.3	1.7	0.7	0.7	8.0	0.0	5.0	5.3	0.3	5.0	0.0	0.0	0.0	0.0	0.3	2.0	1.3	0.0	8.3	0.4235	20.3	0.4	1.0	moderately sorted

Table A.1 (continued)

	Sample	Detrital components	Quartz	Feldspar	RF	Mica	Matrix	Mn. Grains and Accessories	Authigenic components	Quartz cement	Fspar cement	Clay cement	Clay rpl	Ill rpl kaol	Carbonate	Carbonate rpl	Sulphate cements	Pyrite	Fracture fill cement	Fe oxides/-hydroxides	Porosity, permeability, and IGV	Secondary porosity (intragranular)	Secondary porosity (intergranular)	Fracture porosity	He porosity unstressed	Permeability at 2 MPa	Intergranular Volume (IGV)	Grainsize	Average	Maximum	Sorting
Well A	A_400761		57.7	0.0	6.0	0.7	0.0	0.0		15.3	0.0	4.7	2.3	0.7	10.0	0.0	0.0	0.0	0.0	0.0		2.3	0.3	0.0	5.9	0.1923	30.3		0.4	0.9	moderately sorted
	A_400927		54.0	0.7	13.3	0.0	0.0	0.7		15.0	0.3	3.7	2.7	1.0	3.7	0.0	0.0	0.0	0.0	0.0		3.7	1.3	0.0	9.1	0.3246	24.0		0.4	0.9	moderately sorted
	A_401040		61.0	0.0	9.7	0.0	0.0	0.0		13.3	0.0	3.0	3.0	2.3	1.7	0.0	0.0	0.0	0.0	0.0		3.7	2.3	0.0	9.5	0.8313	20.3		0.4	1.0	moderately sorted
	A_401298		64.7	0.3	11.7	0.3	0.0	0.0		6.7	0.0	3.0	4.0	0.3	3.3	0.0	0.0	0.0	0.0	0.0		5.0	0.7	0.0	9.0	0.3760	13.7		0.4	0.7	moderately sorted
	A_401399		55.3	0.3	14.7	1.0	0.0	0.0		13.0	0.0	2.3	3.7	0.7	4.7	0.0	0.0	0.0	0.0	0.0		2.7	1.7	0.0	8.4	0.2131	21.7		0.3	0.6	moderately sorted
	A_414348		54.7	0.0	5.3	0.0	0.0	0.3		18.0	0.0	0.7	6.0	0.0	4.7	0.0	0.3	0.0	0.0	9.0		0.3	0.7	0.0	6.1	0.1362	33.3		0.2	0.6	moderately sorted
	A_414568		59.7	0.3	6.0	0.7	0.0	0.0		9.0	0.0	0.0	1.3	0.0	0.7	0.0	6.7	0.0	0.0	15.7		0.0	0.0	0.0	4.7	0.1939	32.0		0.2	0.5	moderately sorted
	A_435318		53.7	0.7	18.0	0.7	0.0	0.3		9.0	0.0	2.3	5.7	0.7	5.0	0.0	0.0	0.0	0.0	0.0		3.7	0.3	0.0	7.5	0.2300	16.7		0.3	0.6	moderately sorted
	A_435392		61.3	0.0	8.3	0.3	0.0	0.3		9.0	0.0	3.0	6.7	0.3	8.0	0.3	0.0	0.0	0.0	0.0		2.3	0.0	0.0	6.8	0.3216	20.0		0.4	1.0	moderately sorted
	A_435913		60.0	0.0	5.3	0.0	0.0	0.0		19.7	0.0	2.3	4.0	0.3	8.0	0.3	0.0	0.0	0.0	0.0		0.0	0.0	0.0	1.8	0.0585	30.0		0.2	0.4	moderately sorted
	A_435985		55.3	0.0	9.7	1.0	0.0	0.0		11.7	0.3	2.0	3.3	0.3	16.0	0.0	0.0	0.0	0.0	0.3		0.0	0.0	0.0	1.8	0.0453	30.3		0.2	0.5	moderately sorted
	A_436656		49.3	0.3	19.7	1.0	0.0	0.0		9.3	0.0	4.0	2.0	1.3	12.3	0.3	0.0	0.0	0.0	0.3		0.0	0.0	0.0	2.8	0.1000	26.0		0.2	0.4	moderately sorted
	A_436744		52.3	1.0	16.3	1.7	0.0	0.0		7.7	0.0	5.0	3.0	1.0	10.7	0.7	0.0	0.0	0.0	0.0		0.3	0.3	0.0	2.6	0.0256	23.7		0.2	1.0	moderately sorted
	A_436895		53.0	0.3	15.0	0.3	0.0	0.0		10.7	0.3	2.7	7.7	0.0	4.7	0.0	0.0	0.0	0.0	1.3		3.0	1.0	0.0	7.3	0.1256	20.7		0.3	1.0	moderately sorted
	A_437073		56.3	0.0	13.3	1.0	0.0	0.0		7.7	0.0	3.0	5.7	0.3	11.0	0.0	0.0	0.0	0.0	0.0		1.3	0.3	0.0	6.0	0.0884	22.0		0.3	0.6	moderately sorted
	A_437383		59.0	0.0	10.7	1.0	0.0	0.7		9.7	0.0	4.7	4.7	0.3	7.0	0.0	0.0	0.0	0.0	0.0		1.7	0.7	0.0	8.3	0.3631	22.0		0.4	2.0	moderately sorted
	A_437430		50.0	0.3	19.0	2.3	0.0	0.7		10.3	0.0	3.0	6.7	0.7	4.7	0.7	0.0	0.0	0.0	0.0		1.0	0.7	0.0	8.5	0.2017	18.7		0.3	0.6	moderately sorted
	A_437719		56.7	0.3	6.0	0.7	0.0	0.0		11.7	0.0	2.7	10.0	0.0	8.3	0.0	0.0	0.0	0.0	0.0		2.7	1.0	0.0	8.3	0.3404	23.7		0.3	0.9	moderately sorted
	Average		**53.4**	**0.4**	**13.7**	**0.9**	**0.1**	**0.1**		**12.4**	**0.1**	**3.2**	**4.0**	**1.5**	**6.5**	**0.2**	**0.2**	**0.0**	**0.0**	**0.8**		**1.8**	**0.6**	**0.0**	**6.4**	**0.2**	**23.8**		**0.3**	**0.7**	
	Max		**64.7**	**1.7**	**20.7**	**3.0**	**1.3**	**1.0**		**20.0**	**0.3**	**6.0**	**10.0**	**4.3**	**23.3**	**0.7**	**6.7**	**0.0**	**0.0**	**15.7**		**5.0**	**2.3**	**0.0**	**9.5**	**0.8**	**33.3**		**0.4**	**2.0**	
	Min		**44.0**	**0.0**	**5.3**	**0.0**	**0.0**	**0.0**		**4.3**	**0.0**	**0.0**	**0.3**	**0.0**	**0.7**	**0.0**	**0.0**	**0.0**	**0.0**	**0.0**		**0.0**	**0.0**	**0.0**	**1.8**	**0.0**	**13.7**		**0.2**	**0.4**	
Well B	B_216082		63	0.7	12.7	2.7	3.3	0.0		6.7	0.0	3.0	1.0	0.3	4.0	0.0	0.3	0.0	0.0	0.0		0.0	2.3	0.0	16.9	29.8703	19.7		0.29	0.62	moderately sorted
	B_216236		56.3	0.0	16.7	9.0	2.7	1.7		2.3	0.0	3.3	1.0	1.0	2.7	0.0	0.0	0.0	0.0	0.0		0.3	3.0	0.0	8.3	0.6804	14.0		0.22	0.61	moderately sorted
	B_225502		56.3	0.3	17.0	2.0	1.3	1.0		7.7	0.0	2.7	0.7	1.3	9.0	0.3	0.0	0.0	0.0	0.0		0.0	0.3	0.0	7.0	0.0199	21.0		0.22	0.46	moderately sorted
	B_226012		57.3	0.0	16.0	1.3	0.7	0.3		9.0	0.0	6.0	0.3	2.3	4.3	0.3	0.0	0.0	0.0	0.0		0.0	2.0	0.0	10.7	0.1654	22.0		0.33	0.95	moderately sorted
	B_227129		57.3	0.0	20.3	1.3	0.3	0.3		7.7	0.0	2.7	1.3	0.7	5.7	0.0	0.0	0.0	0.0	0.0		0.0	2.3	0.0	10.0	0.1654	18.7		0.39	4.56	moderately sorted
	B_229614		64.3	0.3	17.0	0.7	0.0	0.3		4.7	0.3	2.7	0.0	1.0	7.3	0.3	0.0	0.0	0.0	0.0		0.0	1.0	0.0	8.6	0.0305	16.0		0.51	6.36	moderately sorted
	B_232413		50.7	0.0	25.0	3.0	0.0	0.3		4.0	0.0	6.3	0.7	1.3	7.7	0.3	0.0	0.0	0.0	0.0		0.3	0.3	0.0	9.6	0.0477	18.3		0.27	0.77	moderately sorted
	B_232596		56.7	0.3	16.3	2.7	0.0	0.0		7.3	0.0	4.0	0.7	1.0	9.7	0.0	0.0	0.0	0.0	0.0		0.0	1.3	0.0	5.1	0.0251	22.3		0.33	1.26	moderately sorted
	B_232740		46.7	0.0	28.0	7.3	0.3	1.7		1.3	0.0	2.3	0.0	1.3	8.3	0.0	0.0	0.0	0.0	0.3		0.7	1.7	0.0	-	-	14.3		0.19	0.42	moderately sorted
	B_233631		60.0	0.3	13.3	2.0	0.0	0.0		7.0	0.0	3.3	0.7	0.7	8.3	0.3	0.0	0.0	0.0	0.7		0.7	2.7	0.0	8.4	0.0477	22.0		0.32	1.46	moderately sorted
	B_233785		55.7	0.7	13.0	1.0	0.0	0.3		8.7	0.3	4.7	0.3	2.0	9.7	0.0	0.3	0.0	0.0	0.7		0.0	2.7	0.0	8.0	0.0199	27.0		0.31	0.99	moderately sorted
	B_234230		69.3	0.0	9.7	1.0	0.0	0.0		5.7	0.0	4.3	0.0	2.0	6.0	0.0	0.0	0.0	0.0	0.3		0.7	1.0	0.0	9.4	0.0361	17.3		0.33	0.99	moderately sorted

Table A.1 (continued)

	Sample	Detrital components	Quartz	Feldspar	RF	Mica	Matrix	Mn. Grains and Accessories	Authigenic components	Quartz cement	Fspar cement	Clay cement	Clay rpl	Ill rpl kaol	Carbonate	Carbonate rpl	Sulphate cements	Pyrite	Fracture fill cement	Fe oxides/-hydroxides	Porosity, permeability, and IGV	Secondary porosity (intragranular)	Secondary porosity (intergranular)	Fracture porosity	He porosity unstressed	Permeability at 2 MPa	Intergranular Volume (IGV)	Grainsize	Average	Maximum	Sorting
Well B	B_234404		62.7	0.3	14.0	3.0	0.0	0.0		5.0	0.0	4.7	0.7	2.0	5.0	0.0	0.0	0.0	0.0	0.3		0.3	2.0	0.0	9.0	0.0149	17.0		0.25	0.57	moderately sorted
	B_234491		62.7	0.0	12.3	1.3	0.0	0.0		7.0	0.0	3.7	0.7	1.3	5.3	0.0	0.3	0.0	0.0	0.7		1.3	3.3	0.0	10.8	0.0784	20.3		0.27	2.09	moderately sorted
	B_234824		62.3	0.7	9.3	0.7	3.0	0.0		2.7	0.0	2.3	0.7	1.7	14.3	0.0	0.0	0.0	0.0	0.0		1.0	1.3	0.0	5.6	5.9245	23.7		0.36	0.82	moderately sorted
	B_237706		63.7	0.0	12.0	0.0	0.0	0.0		3.0	0.0	3.0	0.3	1.3	12.7	0.0	0.0	0.0	0.0	0.0		0.7	3.3	0.0	10.8	0.1723	22.0		0.31	0.70	moderately sorted
	B_241160		53.3	0.7	17.7	0.7	0.0	0.0		6.0	0.0	3.7	1.7	1.7	0.7	0.0	0.0	0.0	0.0	10.7		1.3	2.0	0.0	7.4	0.7521	23.0		0.17	0.44	moderately sorted
	B_241228		51.7	0.0	16.7	3.7	0.0	0.3		5.0	0.0	2.7	1.0	2.3	3.7	0.0	0.0	0.0	0.0	10.3		0.7	2.0	0.0	8.2	0.0251	23.7		0.16	0.39	moderately sorted
	B_241520		56.7	0.0	13.3	0.3	0.0	0.0		12.0	0.3	1.7	2.3	2.7	4.0	0.0	0.0	0.0	0.0	5.0		1.0	0.7	0.0	-	-	23.7		0.28	0.77	moderately sorted
	B_242098		57.7	0.0	17.0	1.7	0.0	0.0		6.3	0.0	3.3	2.0	3.7	5.0	0.0	0.0	0.0	0.0	0.3		1.0	2.0	0.0	8.4	0.0361	17.0		0.32	1.15	moderately sorted
	B_244485		58.0	0.0	10.7	1.0	0.0	0.0		10.3	0.0	5.7	1.3	4.0	5.3	0.0	0.0	0.0	0.0	0.3		1.3	2.0	0.0	10.5	0.0305	23.7		0.31	1.06	moderately sorted
	B_245525		59.0	0.0	10.3	2.0	0.0	0.0		7.0	0.0	5.3	3.0	5.3	5.7	0.0	0.0	0.0	0.0	0.3		0.3	1.7	0.0	10.4	0.0477	20.0		0.32	1.05	moderately sorted
	B_245606		53.0	0.3	21.0	1.3	0.0	1.7		3.3	0.0	5.3	0.7	3.3	6.7	0.0	0.0	0.0	0.0	2.3		0.3	0.7	0.0	8.5	0.0102	18.3		0.23	0.66	moderately sorted
	B_245903		54.3	0.3	21.0	1.7	0.0	1.0		5.7	0.0	5.7	1.0	2.7	1.0	0.0	0.0	0.0	0.0	4.7		0.7	0.3	0.0	7.0	0.1243	17.3		0.14	0.28	moderately sorted
	B_246091		49.3	0.0	17.3	2.3	0.0	1.0		8.3	0.0	6.7	1.3	6.0	4.7	0.0	0.0	0.0	0.0	0.0		2.0	1.0	0.0	11.4	0.0477	20.7		0.24	0.64	moderately sorted
	B_247389		55.3	0.7	17.7	1.0	0.0	0.0		8.0	0.0	5.0	0.7	6.0	2.3	0.0	0.0	0.0	0.0	0.0		1.3	2.0	0.0	10.8	0.0721	17.3		0.23	0.62	moderately sorted
	B_247495		68.3	0.0	7.0	0.3	0.0	0.3		8.7	0.0	2.3	2.0	5.0	4.0	0.0	0.0	0.0	0.0	0.0		0.3	1.7	0.0	9.5	0.1723	16.7		0.42	2.21	moderately well sorted
	B_248360		59.3	0.3	18.3	1.7	0.0	0.7		3.7	0.0	6.7	2.7	4.0	1.3	0.0	0.0	0.0	0.0	0.0		0.3	1.0	0.0	10.2	0.2005	12.7		0.17	0.50	moderately sorted
	B_252025		51.7	0.0	18.0	3.0	0.0	0.3		2.3	0.3	6.7	5.3	4.7	6.0	0.0	0.0	0.0	0.0	1.3		0.3	0.0	0.0	7.6	0.2148	16.7		0.20	0.47	moderately sorted
	B_252713		58.7	0.0	11.0	0.3	0.0	0.7		8.0	0.0	4.3	3.7	8.0	2.0	0.3	0.0	0.0	0.0	0.0		2.7	0.3	0.0	-	-	14.7		0.24	0.69	moderately sorted
	B_255872		57.0	0.3	14.7	0.3	0.0	0.0		5.0	0.0	4.0	6.3	5.3	5.3	0.0	0.0	0.0	0.0	0.0		0.0	1.7	0.0	6.1	40.4699	16.0		0.28	0.88	moderately sorted
	B_258134		55.7	1.0	14.0	1.3	0.0	0.7		7.3	0.0	3.3	2.7	3.7	8.0	0.3	0.0	0.0	0.0	0.0		1.3	0.7	0.0	12.0	0.0960	19.3		0.24	0.63	moderately sorted
	B_258160		60.0	0.7	17.3	3.0	0.0	0.7		9.0	0.0	1.0	2.0	1.3	1.7	0.3	0.3	0.0	0.0	0.0		1.7	0.3	0.7	13.7	0.1109	12.3		0.18	0.41	moderately sorted
	B_258209		52.0	0.7	15.0	1.0	0.0	0.7		6.3	0.0	2.7	7.3	4.7	4.3	0.7	0.0	0.0	0.0	1.0		2.7	1.0	0.0	13.5	0.0848	15.3		0.23	0.47	moderately sorted
	B_258241		64.0	0.3	13.3	1.0	0.0	0.0		10.0	0.3	2.7	0.7	3.3	3.3	0.0	0.0	0.0	0.0	0.3		0.7	0.0	0.0	11.5	0.0550	16.7		0.29	1.14	moderately sorted
	B_258367		65.7	0.3	15.0	0.3	0.0	0.3		4.7	0.0	2.0	2.0	4.0	3.0	0.3	0.0	0.0	0.0	0.0		1.7	0.7	0.0	13.2	0.1320	10.3		0.22	0.82	moderately sorted
	B_258438		59.7	0.3	14.3	0.7	0.0	0.0		3.3	0.0	1.7	1.7	1.0	16.7	0.0	0.0	0.0	0.0	0.0		0.7	0.0	0.0	6.1	0.0380	21.7		0.27	0.61	moderately sorted
	B_259518		59.3	0.3	16.7	2.0	0.0	0.3		5.0	0.0	3.3	3.7	5.0	0.7	0.0	0.0	0.0	0.0	0.3		2.0	1.3	0.0	9.1	0.0305	10.7		0.18	0.52	moderately sorted
	B_259755		67.0	0.0	10.0	0.3	0.0	0.3		7.7	0.0	1.3	1.7	2.0	6.7	0.3	0.0	0.0	0.0	0.3		1.3	1.0	0.0	11.7	0.0580	17.0		0.29	0.74	moderately sorted
	B_260295		71.0	0.0	10.0	0.0	0.0	0.0		9.3	0.0	2.0	0.7	3.7	1.7	0.0	0.0	0.0	0.0	0.0		1.0	0.7	0.0	11.9	0.0000	13.7		0.33	3.25	moderately sorted
	B_260795		66.7	0.0	6.0	0.0	0.0	0.0		12.0	0.0	1.0	1.3	5.0	4.3	0.0	0.0	0.0	0.0	0.0		2.7	1.0	0.0	11.4	0.0560	18.3		0.30	0.92	moderately sorted
	B_261178		52.0	0.7	24.7	1.3	0.3	0.7		4.3	0.0	3.3	2.7	5.0	1.0	0.3	0.0	0.3	0.0	0.3		2.0	1.0	0.0	10.7	0.1043	10.7		0.20	1.06	moderately sorted
	B_261657		60.7	0.7	8.0	0.7	0.0	0.3		10.3	0.0	3.7	2.3	4.7	3.0	0.0	0.0	0.0	0.0	0.0		3.0	2.7	0.0	13.7	0.0305	19.7		0.39	1.28	moderately sorted
	B_273810		65.3	0.3	15.3	0.7	0.0	0.3		3.0	0.0	5.3	1.7	1.3	5.7	0.0	0.0	0.0	0.0	0.0		0.3	0.7	0.0	-	-	14.7		0.22	0.52	moderately sorted
	B_273940		61.3	0.0	12.0	1.0	0.0	0.3		8.7	0.0	2.3	3.7	0.7	8.0	0.3	0.0	0.0	0.7	0.0		0.3	0.7	0.0	-	-	19.7		0.29	1.05	moderately sorted

Table A.1 (continued)

Well B

Sample	Detrital components	Quartz	Feldspar	RF	Mica	Matrix	Mn. Grains and Accessories	Authigenic components	Quartz cement	Fspar cement	Clay cement	Clay rpl	Ill rpl kaol	Carbonate	Carbonate rpl	Sulphate cements	Pyrite	Fracture fill cement	Fe oxides/-hydroxides	Porosity, permeability, and IGV	Secondary porosity (intragranular)	Secondary porosity (intergranular)	Fracture porosity	He porosity unstressed	Permeability at 2 MPa	Intergranular Volume (IGV)	Grainsize	Average	Maximum	Sorting
B_273945		61.3	0.0	13.0	0.3	0.7	0.3		11.7	0.3	2.3	1.7	1.3	3.7	0.0	0.0	0.0	0.0	0.3		1.3	1.7	0.0	-	-	20.7		0.23	0.60	moderately sorted
B_282710		52.3	0.3	17.3	2.7	0.0	0.0		7.7	0.0	4.3	5.3	3.3	4.7	0.3	0.0	0.0	0.0	0.0		1.0	0.7	0.0	7.8	0.0149	17.3		0.23	0.56	moderately sorted
B_282784		48.7	1.0	26.0	6.7	0.0	1.0		5.0	0.3	2.0	0.3	2.0	3.7	0.0	0.0	0.0	1.7	0.7		0.7	0.3	0.0	3.9	0.0059	12.0		0.12	0.35	moderately sorted
B_283172		58.3	1.0	17.0	2.7	0.0	0.3		5.7	0.0	2.0	2.0	3.3	4.0	0.0	0.0	0.0	0.0	0.3		1.7	1.7	0.0	8.0	0.1043	13.7		0.18	0.39	moderately sorted
B_289835		54.0	0.3	14.0	0.0	0.0	0.0		7.7	0.0	2.0	1.3	0.7	18.3	0.3	0.0	0.0	0.0	0.7		0.3	0.3	0.0	-	-	29.0		0.19	0.49	moderately sorted
B_289885		53.7	0.0	13.3	0.3	0.0	0.3		6.7	0.0	2.7	5.3	1.0	15.3	0.3	0.0	0.3	0.0	0.3		0.3	0.0	0.0	-	-	25.3		0.18	0.56	moderately sorted
Average		**58.4**	**0.3**	**15.2**	**1.7**	**0.2**	**0.4**		**6.6**	**0.0**	**3.5**	**1.8**	**2.8**	**5.8**	**0.1**	**0.0**	**0.0**	**0.0**	**0.8**		**0.9**	**1.3**	**0.0**	**9.6**	**1.9**	**18.3**		**0.3**	**1.0**	
Max		**71.0**	**1.0**	**28.0**	**9.0**	**3.3**	**1.7**	**0.0**	**12.0**	**0.3**	**6.7**	**7.3**	**8.0**	**18.3**	**0.7**	**0.3**	**0.3**	**1.7**	**10.7**	**0.0**	**3.0**	**3.3**	**0.7**	**16.9**	**40.5**	**29.0**		**0.5**	**6.4**	
Min		**46.7**	**0.0**	**6.0**	**0.0**	**0.0**	**0.0**	**0.0**	**1.3**	**0.0**	**1.0**	**0.0**	**0.3**	**0.7**	**0.0**	**0.0**	**0.0**	**0.0**	**0.0**	**0.0**	**0.0**	**0.0**	**0.0**	**3.9**	**0.0**	**10.3**		**0.1**	**0.3**	